Concrete and Mortar Production Using Stone Siftings

Concrete and Mortar Production Using Stone Siftings

L. Dvorkin, V. Zhitkovsky
National University of Water and Environmental Engineering
Rivne, Ukraine

Y. Ribakov
Ariel University
Israel

CRC Press
Taylor & Francis Group
Boca Raton London New York

CRC Press is an imprint of the
Taylor & Francis Group, an **informa** business

A SCIENCE PUBLISHERS BOOK

CRC Press
Taylor & Francis Group
6000 Broken Sound Parkway NW, Suite 300
Boca Raton, FL 33487-2742

First issued in paperback 2020

© 2018 by Taylor & Francis Group, LLC
CRC Press is an imprint of Taylor & Francis Group, an Informa business

No claim to original U.S. Government works

ISBN-13: 978-1-138-56558-6 (hbk)
ISBN-13: 978-0-367-78136-1 (pbk)

Library of Congress Cataloging-in-Publication Data

Names: Dvorkin, L. I. (Leonid Iosifovich), author. | Zhitkovsky, V., author. | Ribakov, Yuri, author.
Title: Concrete and mortar production using stone siftings / L. Dvorkin, National University of Water and Environmental Engineering, Rivne, Ukraine, V. Zhitkovsky, National University of Water and Environmental Engineering, Rivne, Ukraine, Y. Ribakov, Ariel University, Israel.
Description: First edition. | Boca Raton, FL : Taylor & Francis Group, CRC Press, [2018] | Includes bibliographical references and index.
Identifiers: LCCN 2018001559 | ISBN 9781138565586 (hardback)
Subjects: LCSH: Aggregates (Building materials) | Concrete--Materials. | Concrete--Additives. | Stone industry and trade--By-products. | Waste products--Recycling. | Stone, Crushed.
Classification: LCC TA441 .D86 2018 | DDC 624.1/832--dc23
LC record available at https://lccn.loc.gov/2018001559

Visit the Taylor & Francis Web site at
http://www.taylorandfrancis.com

and the CRC Press Web site at
http://www.crcpress.com

Preface

The monograph analyses the state of the problem in using stone siftings and aspiration dust obtained in natural stone crushing for producing concrete aggregates and fillers for dry construction mixtures and mortars on their basis. The influence of disperse fraction in stone siftings and aspiration dust on structural, mechanical and rheological properties of cement composite construction materials is investigated. Hypothesis for obtaining technological conditions, providing positive effect of the disperse fraction on strength and other properties of cement based concrete and mortar is proposed.

Experimental results on studying properties of dry mixtures and mortars on their basis using stone crushing aspiration dust as filler are presented. Efficiency of using fillers, based on igneous rocks, on adhesive and other properties of mortars is demonstrated. Methodology for design of mortars composition for given mortar properties in case, when aspiration dust is used as filler, is proposed.

The monograph presents experimental results on fine-grained concrete including as a main aggregate stone siftings with up to 20% of disperse fraction. It is shown that it is possible to produce fine grain concrete class C20/25 … C60/75. Technological parameters of vibro-pressed fine-grained concrete with raw stone siftings are developed. Methodologies for composition design of fine-grained concrete with given workability are proposed. Possibility for producing macroporous light-weight concrete for walls and filtration materials, based on stone siftings fillers is shown. The authors express their sincere thanks to the colleagues from the department of construction materials and materials sciences who participated in the experimental programme that are used in the monograph. The authors hope that the monograph will be interesting for experts and specialists working in the field of construction materials technology. The authors would be thankful for all comments and recommendations regarding the book.

Contents

CHAPTER 1

Problem of Using Industrial Waste

Wastes of Stone Crushing

1.1 Problem of using industrial wastes in construction

Industrial production is increasing annually globally and in proportion to its growth, the amount of waste also increases by about 2 times, every 8 ... 10 years. A rational solution to the industrial waste utilization problem depends on a number of factors: quantity, material composition and aggregate state of waste, features of the technological cycle and waste extraction from the production, etc. Reduction of losses, caused by formation of industrial waste, can be achieved by improving production technologies, increasing the efficiency of waste extraction, disposal, and rational storage. The most effective solution to the industrial waste problem is in applying non-waste technologies [1].

An integrated approach in dealing with this issue is to use the raw materials, industrial waste or by-products of some industries as the source materials of others. At present, such use of raw materials arise due to the needs in the development of various industries. The importance of integrated raw materials processing can be considered in several aspects. First, utilization of waste enables to solve environmental protection problems, to release valuable land used for dumps and sludge [2, 3], to eliminate harmful emissions into the environment Secondly, industrial waste largely satisfy the needs of some processing industries in raw materials, and in many cases it is already somewhat prepared, since they are subjected to the primary production process (grinding, firing, etc.). Thirdly, complex use of raw materials reduces the specific expenditures per unit of output as well as the payback period. The main production costs related to waste storage, construction and maintenance of their storage are also reduced. The heat and electricity consumption for new products decrease and the equipment productivity increases.

One of the most important areas in the consumption of industrial waste and by-products is in the production of construction materials. Given that the cost of material resources is more than 50% of the construction materials, the use of waste is a way to improve the efficiency of construction materials production [4–8].

Wastes are usually classified [9]:

- by aggregate state: solid, liquid and gaseous;
- by chemical composition: organic and inorganic;
- by origin: industrial, agricultural, household.

By aggregate state some types of waste can be in the transition state, i.e., in the state of dispersions, aerosols, emulsions, etc.

The most important constituent of construction materials production are mineral products, which represent most of the waste produced by extractive and processing industries. These products are more explored and more widely used than organic [10, 11].

It is possible to classify industrial by-products at the time of their separation from the main technological process into the following three classes: (A) products that have not lost their natural properties; (B) artificial products obtained as a result of deep physical and chemical processes; (C) products formed during long storage in dumps.

Class A products (career residues and residues after enrichment on minerals) have chemical-mineralogical compositions and properties of the corresponding rocks. Their application range depends on the aggregate state, fractional and chemical composition, physical and mechanical properties. Class A mineral products are mainly used as concrete aggregates, as well as the initial clay, carbonate or silicate raw materials for production of various artificial construction materials (ceramics, lime, autoclave materials).

Class B products are obtained as a result of physical and chemical processes occurring at normal or more often high temperatures. The range of their possible use is wider than that of class A products. It is especially effective to use these wastes in cases allowing full or partial compensation of the fuel and energy resources expenses as well as man power. The use of this class of products is rational in the production of cements, autoclave hardening materials, where the increased reactivity of raw materials yields a high economic effect. Thus, when using blast furnace slag for blast-furnace cement production, the fuel and energy costs per production unit are reduced by almost twice, and the price decreases by 25 ... 30%.

Class C products are formed as a result of physical and chemical processes occurring in dumps (self-ignition, slags decomposition, powder formation, etc.). Typical representatives of raw materials of this class are burned aluminosilicate rocks.

The classification above requires mandatory consideration of the by-products chemical characteristics. Depending on chemical compounds prevailing in their composition, mineral waste can be divided into the following groups: silicate, carbonate, limestone, gypsum and ferruginous. Within each group, a more detailed classification is possible. For example, silicate waste can be subdivided into basic and acid, depending on the percentage of basic and acid oxides, carbonates—on calcium and magnesium. In some cases, for chemical characteristics, the leading place is given to substances contained in relatively small quantities, but have a decisive influence on selecting the proper method for using the waste (for example, alkali, zinc, aluminum-containing compounds).

Most of the natural and manmade industrial mineral wastes consist of silica, silicates and aluminosilicates of calcium and magnesium. This is because they are waste of extraction and processing natural silicate materials, which are 86.5% of the earth crust mass.

Silica waste of industry can be divided into four groups depending on the structure and chemical composition. The first group includes mineral raw materials, in which silica is bound to silicates or aluminosilicates of calcium, which are predominantly in the vitreous state. They acquire hydraulic activity at alkaline and sulfate activation. Depending on the content of CaO and Al_2O_3, such materials harden in normal conditions or under heat processing. At high-temperature by firing them together with calcium carbonate it is possible to get Portland cement clinker. Typical representatives of this group are granulated blast furnaces and phosphorous slags, as well as fuel slag formed at lumpy coal combusting.

The second group includes wastes in which silica is bound to silicates or to aluminosilicates that are in a crystalline state. They do not exhibit activity under normal temperature-humid conditions. This group includes, for example, slowly cooled dump metallurgical slags and electrothermophosphorous slags, as well as by-products of the mining industry.

In the third group waste silica is mainly included in a free state in the form of quartz. Representatives of this group are silicate products—by-products of enrichment of various ores, raw materials for the chemical industry, overburden rocks. Waste of the second and third groups can be used in construction as non-metallic construction materials, raw materials for production of autoclave products ceramics and glass.

The fourth group includes raw materials containing mainly silicates of calcium both in unhydrated and in hydrated state, for example, slag of metallurgical productions (nepheline, bauxite, sulfate, white, etc.). These by-products are used for producing so-called slurry cements, Portland cement and autoclave hardening products.

According to the standards, all industrial wastes are divided into four classes by their hazard (Table 1.1).

Table 1.1 Classification of industrial waste according to their hazard.

Class	Characteristics of substances (waste)
First	Extremely dangerous
Second	Highly dangerous
Third	Moderately dangerous
Fourth	Not so dangerous

Examples of hazard class for some chemicals:

- presence of mercury wastes, sulamites, potassium chromic, antimony trichloride, benzopyrene, arsenic oxide and other highly toxic substances allows them to be classified into the first hazard class;
- presence of copper, nickel, antimony oxide, lead nitrogen and other less toxic substances in waste products makes it possible to transfer these waste to the second hazard class;
- presence of copper sulfate, copper oxalate, nickel chloride, lead oxide, carbon chloride in waste products allows them to be classified as hazardous corresponding to the third class;
- if waste includes manganese sulphate, zinc sulphate or phosphates, zinc chloride it belongs to the fourth hazard class.

In assessing industrial waste as raw materials for construction materials production, it is necessary to take into account their compliance with the norms for the admissible radionuclides content. Both natural and technogenic raw materials may contain radionuclides (radium-226, thorium-232, potassium-40, etc.) that are sources of γ-radiation. At decay of radium-226, radioactive gas is released into the environment. According to expert estimates, it is up to 80% of the total radiation dose of people. Following the European standards, it is forbidden to use materials with radiation above 25 nKi/kg in construction. It is recommended to control materials with radiation from 10 to 25 nKi/kg and consider non-radioactive materials with radiation less than 10 nKi/kg.

1.2 Selecting the way of industrial waste use

Selecting the way of using waste as man-made raw materials is aimed at achieving maximum resource savings, energy consumption reduction and implementing environment friendly technologies [12]. When deciding on the possibility of using technogenic products, it is necessary to use a system of criteria, taking into account the possible scope of their application. The ecological assessment should take into account the data on concentration of heavy metals, toxic substances and activity of natural radionuclides. At low content of heavy metals it is allowed to use waste at high-temperature technologies if the amount of formed melt is sufficient to preserve dangerous substances. Environmentally hazardous wastes cannot be used and sent for disposal before preliminary purification.

Among the most important technical and economic indicators, taken into account when choosing the way of waste disposal, are: the degree of their possible use; saving natural raw materials and resources like fuel, energy; qualitative indicators of products, their demand and competitiveness in the market; homogeneity of the waste composition, ability to technological preparation and processing; availability of transport communications and transportation distance.

One of the main criteria for selecting the way of using the industrial waste is the achieved economic effect.

The economic effect E_{spec} of using 1 t of solid waste in the production of construction materials is defined as the difference between the total specific costs for production of similar materials from traditional raw materials and the operation of dumps and expenses for production of similar materials from industrial waste (or waste of municipal economy):

$$E_{spec} = (n_1/a)(C_1 + n_2 C_2 - C_3) + E_N (K_1 + n_2 K_2 - K_3) \qquad (1.1)$$

where C_1 and C_3 are the cost of construction materials produced using traditional raw materials and raw materials to be utilized, respectively; C_2 is the annual cost of dumps and transportation of by-products; n_1 is a coefficient, taking into account the part of expenses for this type of material in the total cost of raw materials and that of materials in the cost of construction; n_2 is a coefficient, taking into account the partial or total elimination of dumps, $n_2 = 0,3...1$; K_1 i K_3 are specific capital investments for producing construction materials from traditional raw materials and those to be utilized accordingly; K_2 are investments required for dumps construction; E_N is normative coefficient of investments return; a is the specific consumption of raw materials to be disposed per unit of production.

For individual enterprises, the economic efficiency of the waste products use is defined as the ratio of profits derived from the use of waste to the investment:

$$E = (C - S)/K \qquad (1.2)$$

where C is the cost of annual output using waste products; S is the price of annual output using waste products; K is the investment in the implementation of organization and technical measures for the waste recycling.

The overall economic efficiency coefficient of individual waste management measures is recommended to be obtained by the ratio of profit to the cost of the corresponding investment (capital and current):

$$\beta = \frac{C - S}{K} \qquad (1.3)$$

The profit indicator, being derived from the current product wholesale price and the production cost, quite fully reflects the results of rational resources use and represents one of the final indicators of the enterprise economic activity.

The enterprise net profit growth (ΔP_0) as a result of integrated waste utilization is determined by the following formula:

$$\Delta P_0 = \sum_{i=1}^{n} \left[C_{0_i} - (S_{0_i} + D_i) \right] A_{0_i} \qquad (1.4)$$

where C_{0i} is the wholesale price of products, made using material resources of type i (i = 1, 2, 3 ... n); S_{0i} is the cost of producing a unit of a similar products made using waste; D_i are the fixed payments to the budget, taking into account discounts for the use of secondary material resources; A_{0i} are additional (more than previously produced) natural volume of sold commercial products made with the use of waste.

The indicator of material content for the enterprise's products at cost M_s set the part of material costs in the value of marketable products or the costs per unit of production, and is determined:

$$M_S = \frac{\sum_{i=1}^{n} M_{E_i}}{\sum_{i=1}^{n} S_{m_i} \cdot \left(A_{0_i} + A_0\right)}$$ (1.5)

where M_{E_i} is the cost of resource of type i; S_{m_i} is the cost of products unit manufactured using the material resources of type i; A_{0i}, A_i are physical volumes of realized commercial products, made from material resources of type i, respectively, with and without use of secondary material resources, respectively

Indicators of material content help to study the dynamics of material costs at the enterprise, depending on the intensity of the secondary material resources use and to implement the integrated use of waste products in the overall system of resource preservation.

Using certain types of waste instead of primary raw materials requires taking into account the degree of primary raw materials interchangeability by waste. Since products from primary (conditional) raw materials may differ in their consumer properties from those made using waste, these features should be considered in a comparable form. Such product properties as strength, reliability, mass, durability, etc., should be taken as a basis for comparison. For this purpose the following equivalence coefficient of consumer properties (l) is introduced:

$$l = \frac{Q_t}{Q_w}$$ (1.6)

where Q_t is the quantity of products from primary (traditional) raw materials, which is equivalent in terms of their properties to the amount of waste products, t; Q_w is the quantity of products from waste or partial use of waste in the form of additives, t.

The equivalence coefficient of consumer properties is obtained based on experimental study of qualitative indicators with the use of comparable products, depending on the conditions of their application. If a particular kind of waste can be used for different types of products, then in order to obtain the optimal product it is recommended to find the cost saving for each case of manufacturing the product using the waste, compared to that when the product is obtained from the conditioned (primary) raw materials.

The enterprise's activity intensity in using secondary material resources in the production process is measured, in addition to indicators of net profit growth and reduction of material consumption, by other production indicators, which depend on the integrated use of waste. These include an increase in manufactured products volumes; growth in return on investments; increase in labor productivity; lower production cost; increase in total profits and production profitability.

More accurate calculation of the economic effect due to the use of industrial waste is possible by taking into account the additional effect of reducing the damage caused to the biosphere, ΔY, since there is no need for waste storage areas ΔY_a, as well as reductions of emission in air and water ΔY_b:

$$\Delta Y = \Delta Y_a + \Delta Y_b \qquad (1.7)$$

where

$$\Delta Y_a = \Delta C I S_a / T_a \qquad (1.8)$$

and C_l is the cost of land unit area; S_a is the land area, released from the waste; T_a is the time, during which dumps are developed, years.

1.3 Waste of stone crushing and ways for its application

The most large-scale natural stone processing industry is the production of crushed stone for cement concrete [13–15]. Most of high-strength aggregates for concrete are produced from intrusive igneous rocks. These deposits are represented by such mountain rocks as granite, granodiorite, diorite, gabbro, diabaz, basalt and others. The classification of igneous rocks takes into account their chemical and mineral composition, as well as structure. There are three groups of igneous rocks: acid, medium and basic. The basic rocks include mainly such minerals as pyroxene, olivine and calcium plagioclase. Large amount of the first two minerals in the rocks determines their dark coloration. Acid stones are light colored and consist mainly of alkaline feldspars, sodium plagioclase and quartz. The composition of medium rocks has intermediate characteristics between acid and basic.

Granite is the most common and commonly used rock for producing crushed stone. Large volumes of crushed stone are also made of basalt and andesite. Quartzite, marbled limestone, and dolomite crushed stones are produced from metamorphic rocks, and from sedimentary rocks, sandstone and limestone are obtained.

Crushed stone production yields about 18 to 25% of siftings of the mass of the processed rocks. Recently the demand for so-called "cubic" crushed stone has significantly increased, while the yield of siftings during its production may reach 30 ... 36% and even more depending on the structural and texture features of the rocks. Volumes of igneous rocks siftings, including those resulted by producing crushed stone with an improved grains form, is constantly growing. For typical enterprises with an average capacity of 350 thousand m^3 of crushed stone, approximately 80 ... 110 thousand m^3 of siftings is formed annually. Increasing the amount of waste leads to formation of dumps that occupy large agricultural land areas, as well as air and water pollution [16–18].

Developing modern technological schemes for producing concrete and mortar aggregates require a number of additional operations for sorting, cleaning and, if necessary, cleaning waste. The waste can be cleaned using dry or wet methods. Dry methods are based on crushing impurities by blow or abrasion in the environment of cold or hot gases and further separation from stone material by screwing, pneumatic methods, etc. Wet methods (Fig. 1.1) provide separation of impurities by wet screwing at the final production stages. Wet sieving is rational to remove impurities that are easily separated. In this case, by using special classifiers pure fraction of sand can be separated from waste.

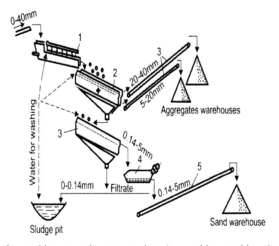

Fig. 1.1 A scheme for crushing screenings processing: 1 – washing mashine; 2 – inertion screens; 3 – screens; 4 – vibration vacuum water eliminator; 5 – conveyors.

Washing crashing waste with contaminants above 10% in conventional equipment is ineffective. An effective washing method is using a vibro-acoustic method based on maximum energy concentration for the disintegration of clay rocks. This method combines low-frequency vibration and acoustic effects. Low-frequency hydro-acoustic devices are used as sources of acoustic vibration. For example, at an average raw material contamination of 12.45%, the vibro-acoustic device reduces pollution to 0.6%. At the same time, when raw materials contamination is 2.7%, trough-type washing can reduce pollution to just 0.76%. Using vibro-acoustic devices enables to process waste with contamination of 40% and even more [19].

Jet type washing machines and vibro-vacuuming water-dispensing devices have been developed and successfully tested in the industry. The latter, due to vacuum suction, reduce the final product moisture content up to 11 ... 13%, allowing its conveyor transportation.

Dry air classification technology of siftings using cascade-gravity classifiers is implemented in practice. It is based on separating fine-grained and sandy materials in the air flow by fineness and particle density due to the interaction of two oppositely directed forces: the gravity acting on the particles of the source material and the ascending air flow. Such machines enable to classify loose materials with a maximum fineness of 10 mm and a moisture content of up to 6% with separation into 2 ... 3 classes, i.e., obtaining 2 ... 3 products.

Since most siftings have a fineness modulus of 3.2 ... 3.6, they are used as an enlargement additive to fine sands in various kinds of cement concrete. Using siftings, it is possible to obtain optimal compositions of concrete and mortars, providing the required construction and technical properties at lowest cost.

Crushing siftings significantly differ in grains surface and shape, mineral and grain compositions (in comparison with natural sands), have higher water demand and voidness, which complicates their use in concrete. The voidness of crushing siftings, consisting mainly of coarse fractions, varies within 40 ... 50%, which is significantly higher than that of natural sands (35 ... 40%).

The main consumers of stone crushing siftings are road construction organizations that use siftings in asphalt mixes as an aggregate. The dusty component of siftings of the main igneous rocks allows partial replacement of carbonate rocks as mineral powder.

Crushing sands are successfully used in highways construction. Siftings of igneous rocks can be used to prepare sandy mixtures treated by cement to be used for drainage, antifreeze layers or for ice sprinkling.

In construction the practice of using dry construction mixtures (DCM) with quartz sand as traditional filler are increasingly being used. However, lack of conventional raw material in a number of regions forces the manufacturers to use local raw materials, including sand from crushing siftings. Sand from various types of rocks is successfully used in various DCMs.

In the US dry mixtures were developed from 20 to 95% of crushed rock by the mixture mass. In Germany plastering DCM compositions were patented that used such rocks as granite and basalt, the grain size of which is from 0.2 to 2.5 mm, which enables to obtain surfaces simulating different rocks. Due to available technology, decorative plastering with the use of siftings allow creating original surface structure. Siftings of igneous and metamorphic rocks, having a decorative color and texture, are used for manufacturing floors and facade plates.

Fine disperse stone materials from rocks can be added into the binder if it is required to provide more durable bonding for repairing reinforced concrete structures.

A DCM which includes crushed granite was developed in France for repairing granite sculptures. DCM with improved adhesive properties were obtained in Germany. These mixtures include fine disperse powders from rocks with a specific surface of 5000 cm^2/g in quantities of 5 ... 95% of the binder weight. Fillers for DCM are received by milling of granite, limestone, marble and other siftings to a specific surface of 2700 ... 2800 cm^2/g. Using decorative finishing mortars with these fillers allows saving 5 to 20% of binder.

Combined binders including gabbro, granodiorite, granite, basalt, porphyryte and limestone milled to specific surface of 4000 ... 5700 cm^2/g were investigated.

The highest strength was obtained for the binder based on the fine-milled gabbro, and the highest adhesion ability was found in cement stone samples with granite filler.

1.4 Properties of stone crushing wastes

Crushing siftings. Crushing siftings can be divided into two groups (Table 1.2):

- Group 1 - siftings obtained from crushing igneous rocks and boulders;
- Group 2 - the siftings obtained from crushing of massive sedimentary rocks (sandstones, limestone, dolomite) and gravel.

According to Table 1.2, siftings of the first group contain less high dispersed impurities compared to the second one.

Siftings of crushing that correspond to the requirements of existing standards can be considered as crushing sand [20, 21]. Sand from crushing siftings is usually obtained directly in the crushed stone production process. Special equipment and appropriate technologies can be used to improve the sand properties. The main part of siftings remains in an unenriched form and accumulates in the dumps.

During the research, about 20 samples were taken from different quarries, which differed slightly in physical properties. The main indicators of the most common of them are given in Table 1.3 and their grain compositions are represented in Fig. 1.2.

Table 1.2 The content of impurities in two groups of siftings.

No.	Indicator	Group 1	Group 2
1	Contents of grains > 0.16 mm, %	5–30	Up to 65
2	Contents of grains < 0.16 mm, %	5–25	Up to 43
3	Amount of dust and clay particles, %	2–15	3–33
4	The content of clay in pieces	Practically absent	0.15–12

Table 1.3 Properties of granite siftings.

No.	Indicator	Standard requirement	Actual value in sample		
			1	2	3
1	Grain composition:				
	Fineness modulus	1.5–4.0	3.24	3.02	3.3
	Content of grains from 5 to 10 mm, %	Up to 15	13.7	2.4	10.8
	Content of grains > 10 mm, %	Up to 0,5	-	-	-
	Content of grains < 0.16 mm, %	Up to 15	15.1	17.,8	20.1
2	Content of dust and clay particles, % (including lump clay)	Up to 3 Up to 0.35	5.7 -	8.4 -	10.2 -
3	The content of organic impurities (Colorimetric test)	Lighter than standard	Lighter than standard		
4	Bulk density, t/m³	-	1.38	1.41	1.42
5	Voidness, %	-	48.5	47.8	47.2

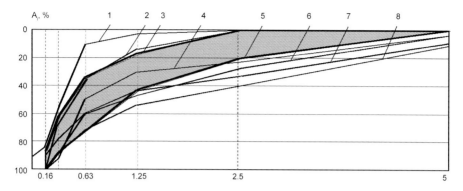

Sieve size of openings, mm

Fig. 1.2 Grain content of aggregates: 1 – quartz sand (M_{fine} = 1.1); 2 – quartz sand, (M_{fine} = 2.0); 3—lower limit of the allowed grain content; 4 – mix of 50% granite siftings and 50% quartz sand (M_f =2.0); 5 – upper limit of allowed grain composition; 6 – mix of 75% granite siftings and 25% of quartz sand (M_{fine} = 2.0); 7 – granite siftings ($m_{0.16}$ =18%); 8 – granite siftings ($m_{0.16}$ = 0%).

All studied siftings were characterized by a significant content of particles < 0.16 mm (15 ... 20%). The content of grains over 5 mm in granite siftings exceeds the standard requirements, according to which for coarse sands of class 1 the content of such grains should be lower than 5% by weight. In order to use sands from sifting in concrete, their grain composition should be optimized, i.e., it should ensure the highest concrete mixture fluidity at minimal cement consumption without concrete mixture stratification [22].

Analysis of the grain composition curves for the studied siftings indicates insufficient content of fraction 0.63–0.315 mm and a significant content of 2.5 ... 10 mm particles. Therefore, they are not suitable for use in cement concrete in accordance with the standards requirements. Adding medium-grained sand (M_f = 2.0) to the siftings can significantly improve their grain composition (Fig. 1.2). Siftings containing about 20% of dust particles are characterized by a lower voidness.

Chemical composition of the investigated granite siftings is given in Table 1.4.

The main chemical component of siftings is silicon oxide. However, it is present in a small amount in the unbound form and preferably in particles less than 0.16 mm. Unlike ordinary construction sands, in granite siftings particles of less

Table 1.4 Chemical composition of the investigated granite siftings.

Sample No.	Content, % by mass											
	SiO_2	TiO_2	Al_2O_3	Fe_2O_3	FeO	MgO	CaO	Na_2O	K_2O	P_2O_5	H_2O	$L.O.I$
1	74.6	0.2	13.9	1.2	0.68	0.5	1.29	3.76	5.29	0.09	0.16	0.7
2	73	0.4	10	0.75	0.58	0.43	1.25	4.12	5.12	0.07	0.13	0.7
3	71	0.2	17	0.95	0.48	0.46	1.32	3.86	5.12	0.08	0.15	0.5
Average	72.9	0.3	13.6	0.97	0.58	0.46	1.29	3.91	5.18	0.08	0.15	0.6

than 0.16 mm contain an increased amount of metal oxides—Al_2O_3, Fe_2O_3. In the investigated crushing siftings the main rock-forming minerals were feldspars, quartz and their fragmentation

The ratio of grains of quartz and feldspar was about 3:1. The mineralogical composition of siftings was quite stable, the variation coefficient of the main rock-forming components—quartz, feldspar and the splices of these minerals did not exceed 30% and in most of the investigated siftings it varied from 10 to 20%. The quartz grains, which have a size preferably in the range of 0.1–0.3 mm, accumulate in the sifting fraction 0.16 ... 0.315 mm and make up 78.4% of this fraction. The fractions of 1.25 ... 2.5 mm and 0.63 ... 1.25 mm were enriched by feldspar grains.

The siftings included mica grains—biotite. In the coarse fractions (up to 0.63 mm) the content of mica was below 2.7%, for fractions 0.315 ... 0.63 mm it was 6%, and for 0.16 ... 0.315 mm increased to 10%. Coarse fractions of siftings (> 5 mm, 2.5–5 mm and 1.25 ... 2.5 mm) include more than 50% of original rock breed debris. A significant part of the stone crushing siftings of igneous rocks with a grain size of 0.16 ... 5.0 mm is formed mainly by corn-shaped grains (79%), and the other part are composed of needle-shaped grains. In fractions from 5 to 1.25 mm, the rough surface of grains predominates, grains with a smooth surface appear in fractions of 1.25 ... 0.63 mm, and among the smaller fractions the number of grains with a smooth surface increases sharply.

Characteristics of particles < 0.16 mm in granite siftings. Specific surface area of particles < 0.16 mm in the studied siftings by Blaine varied in the range from 220 to 240 m^2/kg. For a clearer understanding of the dust grain composition in granite siftings their sedimentation analysis was carried out. The experiment was carried out using torsion scales

Analysis of integral and differential distribution curves of particles by radius (Fig. 1.3) enables to assume that their granulometric composition is non-uniform and interrupted: about 55% of granite dust is represented by particles from 0.16 to 0.13 mm, and the size increases proportionally to percentage content; 15% are particles of 0.13–0.11 mm, the other 30% are particles smaller than 0.11 mm, and the content of particles that differ in size by 0.01 mm is 2.2–3%.

To determine the mineral-petrographic composition of granite siftings particles < 0.16 mm, X-ray analysis on a Dron-4-07 diffractometer was carried out. Analysis of powder-like samples was carried out in continuous mode using Cu Ka radiation and a Ni-filter at a speed of 2°/min.

X-rays diagram (Fig. 1.4) shows that in the composition of granite crushing siftings particles < 0.16 mm dominate feldspars (orthoclase, microcline and albite). Particles of crushing siftings < 0.16 mm have in their composition also clay minerals. According to the X-ray diffraction analysis, particles of granite crushing siftings < 0.16 mm consist of quartz SiO_2 (35%) and feldspars (65%): (orthoclase (~ 30%), microcline (~ 20%), albite (~ 15%) and have a small amount of kaolinite impurities. Based on the available data it is possible to suggest that granite particles < 0.16 mm will actively participate in the hydration and structural formation processes of cement concrete and mortar. As known, fine particles of feldspars serve as an active substrate for the formation of low-calcium hydrosilicates. Depending on the hydration conditions at interaction of feldspars with $Ca(OH)_2$ low-basic calcium hydrosilicates can be formed. At 95°C such interaction is also accompanied by formation of hydrogarnet. With increasing particle dispersion, their activity increases.

Radionuclide analysis of granite siftings was carried out by the enterprises-manufacturers using a gamma-spectrometer at laboratories accredited for radiation control (Table 1.5). According to the criteria for taking decisions on using construction materials, the studied crushing siftings belong to Class I. Results of radionuclide analysis indicate that the investigated siftings are safe for the environment and human health.

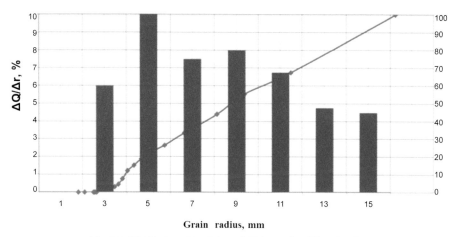

Fig. 1.3 Distribution of particles < 0.16 in granite siftings by size.

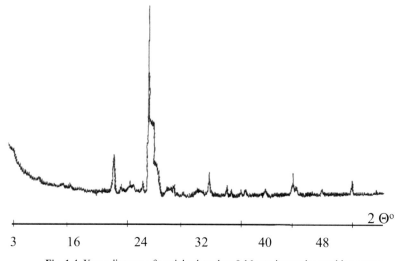

Fig. 1.4 X-ray diagram of particles less than 0.16 mm in granite crushing waste.

Table 1.5 Results of radiation control of granite crushing siftings.

Radioactive element	Specific activity, Bq/kg
Ra-226	35
Th-232	35
K-40	971
A_{ef}	168

Aspiration granite dust. Aspiration granite dust (AGD) is a product caught by aspiration systems when crushing rocks into crushed stone. Depending on the dust collection system type, AGD can be conditionally divided into two types:

• high dispersion aspiration dust (HDAD) caught by the bag filters;
• medium-dispersion aspiration dust (MDAD), caught by cyclones.

Chemical composition of the investigated AGD and its physical characteristics are given in Tables 1.6 and 1.7

Aspiration dust is more dispersed in comparison with siftings particles with dispersity < 0.16 mm, 95% of its particles have dimensions from 0 to $8 \cdot 10^{-9}$ m. The AGD grain composition is shown in Fig. 1.5. By the mineralogical composition the AGD is close to the granite siftings particles < 0.16 mm.

Table 1.6 Chemical composition of aspiration dust.

Indicator	Indicator marking	Quantitative value, %
Calcium oxide	CaO	1.29
Silicon oxide	SiO_2	72.97
Aluminum oxide	Al_2O_3	13.6
Iron oxide(III)	Fe_2O_3	0.98
Iron oxide(II)	FeO	0.58
Magnesium oxide	MgO	0.46
Titanium oxide	TiO_2	0.3
Sodium oxide	Na_2O	3.91
Potassium oxide	K_2O	5.18
Phosphorus oxide	P_2O_5	0.08
Hydrogen oxide	H_2O	0.15
Loss on ignition	LI	0.6

Table 1.7 Physical properties of AGD.

No.	Properties	Value
1	Specific surface, m²/kg - HDAD - MDAD	680 … 720 240 … 260
2	Bulk density, kg/m³ - HDAD - MDAD	590 … 650 820 … 870
3	Grain composition, % up to 20 μm 20…40 μm 40…80 μm 80…160 μm	15,5 … 25,5 36,5 … 45,5 24,9 … 33,5 3,1 … 14,5
4	Content of clay particles, %	0,2 … 0,5
5	Humidity, %	0,3 … 0,5
6	Presence of water-soluble impurities %	no

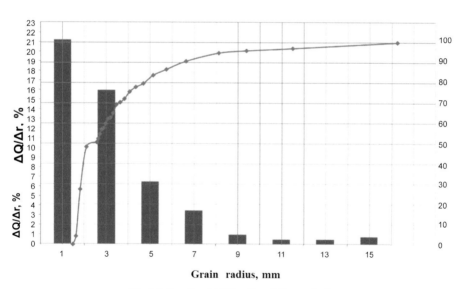

Fig. 1.5 Results of AGD sedimentation analysis.

CHAPTER 2

Theoretical Provisions for Efficient Application of Stone Crushing Wastes in Concrete and Mortars

2.1 Influence of dispersed mineral fillers on the structure formation of cement systems

The main feature of unenriched siftings of rock crushed into stone, which should be taken into consideration when taking decisions regarding the possibility of their use in concrete and mortars, is the high content of the fine dispersive fraction (< 0.16 mm). In aspiration dust this fraction is the main one. By increasing the concrete mixture water demand, this fraction under certain conditions can carry out a positive role. The dust fraction of rocks in concrete and mortars may act as a disperse filler [18, 23–25].

The idea of adding dispersed mineral fillers (microfillers) to cementitious systems is associated with classical research results on the processes of cement stone hydration and structural formation. According to these results, fillers are dispersed particles which do not create their own fields of deformation and stress in the matrix material that allows them to participate in the structure organization of concrete and mortars [26–29].

The expediency of adding fillers in cement composites follows mainly from well-known cement stone representations as a "micro-concrete" in which the hydrated phase, obtained by the chemical interaction of cement grains of less than 30 microns with water, performs the role of a matrix and the role of the filler belongs to coarse cement grains, hydrated from the surface [30, 31]. Based on this concept, by providing a rather high fineness of the clinker component grinding, a part of the cement grains can be replaced with dispersed materials, including those that do not interact chemically with water. Already in the early 30s of the last century cement with carbonate fillers was proposed. It has been found that the fillers in a certain amount do not cause a sharp decrease in strength. The researchers have associated positive effect of the fillers with the so-called "fine powders effect", extending the free space in which the cement hydration products settle down, yielding the hardening acceleration processes.

The expediency of adding fillers to concrete and mortars composition follows, also from the need to ensure their sufficient density without excessive cement consumption. More than 60% of the binder can be used to fill the voids between the aggregate grains, while only a smaller part is actively involved in the adhesive layer formation. Getting a mixture of fillers with minimal voidness is a complex technical and economic task, it requires using many fractions. A more promising technique is increasing the total of content of the powdered part in cement composites due to cement dilution by mineral fillers, introduction of which is possible for both the process of obtaining cement, as well as producing ready for use fillers or in dry mortars and concrete mixtures.

Mineral fillers in cement systems are conventionally divided into active and inert. Active fillers include dispersed minerals that are able to interact with calcium hydroxide released by clinker minerals hydrolysis (so-called active mineral additives). Active fillers also include carbonate powders that interact with the aluminate phase of cement, forming complex compounds such as hydrocarboaluminates. Following modern theoretical representations on the cement systems structure formation, the second group of mineral fillers, which do not directly interact chemically with cement hydration products, is not exactly called inert. They also actively affect the physical and chemical processes of mortars and concretes structure formation and their properties [32, 33].

A more suitable classification of mineral fillers is depending on the degree of their activity relative to cement hydration products. According to this classification, the fillers are divided into 4 groups [18]:

- Group 1 – slowly hardening disperse materials (basic slag, ash, etc.).
- Group 2 – mineral powders that interact chemically with cement hydration products (acid ash and slag, microsilica, metacaoline, etc.).
- Group 3 – disperse materials, characterized by low reactivity at a specific surface of 400 ... 500 m^2/kg (andesites, sienites diabazas, granites).
- Group 4 – chemical inactive fillers (pyrolusites, some metal oxides, etc.).

This classification is also conditional and fair in the case of rocks dispersion up to 600 m^2/kg. It also takes into account only the chemical activity without considering an important filler activity indicator such as its surface energy, which determines the strength of the adhesive contacts in the cement-filler system. In accordance with the thermodynamic concept of adhesion, for strong adhesive contacts formation the filler surface energy should be greater than that of cement.

The influence of fillers in cement systems is evident on the micro, meso and macro levels.

At the micro level, the molecular interaction of cement hydration products with the filler takes place. Chemically active silica fillers as a result of reaction with Ca (OH)$_2$ —hydrolysis product of clinker minerals—form additional amounts of low-calcium hydrosilicates. Carbonate fillers form complex compounds with hydroaluminates and interact with calcium hydrosilicates. The essence of physico-chemical interaction of fillers with hydrated cement consists in the formation of epitaxial contacts, as well as crystallization centers. The latter, according to the Gibbs-Folmer's theory, greatly reduces the required energy for the formation of nucleous of crystals in hardening cement paste. This effect increases at heat treatment, as well as with a decrease in the filler grains radius, adding hardening accelerators. At optimal filler concentration and dispersion a fine-grained cement stone structure is formed, which positively affects its properties.

Under certain conditions, the destructive effect of filler is possible.

It can occur if the filling parameters are over the optimum limits and tensile stresses appear in the whole cement stone volume or in its individual sections.

On the meso level, particles of the filler interact with both the particles of hydrating cement, and with each other. As a result of increasing the solid phase volumetric concentration, part of the mixing water becomes a film state and creates a so-called "compressed conditions", which positively affects the cement stone structure formation.

With a fairly thin layer of a dispersion medium, the filler particles interact with each other. The particles forces adhesion can be found as:

$$F_c = \frac{2}{3}\pi Br / H^3 \tag{2.1}$$

where r is the radius of particles, B is molecular interaction constant and H is the distance between the particles.

The average distance between the filler particles is determined by the ratio between the average volume of the matrix V_m and the total surface of the filler particles:

$$H_{av} = V_m/(V_f S_{s.f}) \tag{2.2}$$

where V_f is the filler volume; $S_{s.f}$ is the specific surface of the filler

Along with surface forces in a cement-water system that contains fillers, Coulomb forces act on the surface at presence of an uncompensated electric charge. Due to different binder and filler nature there is an additional force of electrostatic attraction (F_e) that can be calculated as:

$$F_e = q^2/(12rH) \tag{2.3}$$

where q is the charge; r is the filler particles radius; H is the distance between them.

The maximum strength of the contacts between the filler particles in the cement matrix is achieved by intensive compacting methods of the filled systems or by using plasticizing surface active substances (SAS). The adhesion forces between the individual filler particles increase when the thickness of the binding layer on them decreases.

At the macro level, filler significantly affects the adhesion strength between cement paste and filler, as well as the cohesive strength of the cement composites matrix component and their density. By reducing the inter-granular distances in concrete and mortars, fillers can significantly increase the contact area strength, which positively affects the strength of the materials as a whole. With decreasing the inter-granular distances in quartz sand mortars from 210 to 30 microns, the cement stone hardness has increased 1.5 ... 2 times. The minimum required thickness of the cement paste layer on the aggregate grains is approximately $13 \cdot 10^{-6}$ m. In order to achieve such a thin cement paste layer, a sufficient amount of filler should be added to the composition.

2.2 Influence of mineral fillers on the properties of cement mortars and concrete

The influence of various types of dispersed mineral fillers on the properties of cement mortars and concretes was studied by many researchers. The obtained results depend on the type of fillers, their hydraulic, pozzolan and surface activity, dispersion, grain composition, method of adding and content.

Properties of mortars and concretes containing mineral fillers depend to a large extent on their water demand, which in turn affects the water consumption of mixtures, required for achieving the necessary workability. Water demand of fillers depends on water absorption, forms and features of the grains surface, dispersion, etc. [16, 34, 35].

Most mineral fillers in one way or another increase the water demand of mortar or concrete mixtures. At the same time, such filler such as fly ash at optimal concentration can cause a plasticization effect due to the glassy surface of the particles and their spherical shape. This effect decreases or disappears completely by applying coarser dispersed ash containing a high amount of un–burnt carbon particles with high water absorption and particles of irregular form. Increasing the water demand of fillers directly depends on the dust fraction content. Increase in the water demand of fillers by 1% yields a corresponding increase in water demand of the concrete mixture by approximately 5 liters.

The filler is held in the cement-water system by coagulation contacts whose strength (f_k) is:

$$f_k = vf(F_p, \varphi, S_{spec}^2) \qquad (2.4)$$

where v is the chemical interaction constant, F_p is the resultant interaction force between the particles, φ is the degree of filling, S_{spec} is the specific surface of particles.

Formation of strong coagulation contacts reduces the water separation of filled cement-water mixtures.

The influence of disperse fillers on cement composites strength is usually positive up to a certain optimal concentration, if a significant increase in the mixture water demand is avoided.

The strength of filled cement systems is formed as a result of the synthesis of chemical, physical–chemical and physical–mechanical processes in the cement-water-filler system. Filler takes an active part in these processes. The nature and magnitude of the effect may vary in a wide range depending on the extent to which the composition and structure of the material change. In this case, such factors as degree of hydration, phase composition of the filled cement stone, cement stone adhesive strength in relation to fillers, crystalline and pore structure peculiarities are important [14, 36–38].

An increase in hydration degree and hydrated volume is the main result of adding fillers with high enough puzzolan activity, such as ash and slag [39–42]. When adding highly dispersed silica fillers with high surface energy, such as microsilica and metaacolin, the conditions for formation of cement stone structure condensation and crystallization are significantly changed. In this case formation of crystals nuclei of hydrated compounds is accelerated, the structure becomes much denser, the adhesive strength and cement stone strength in the contact zone with fillers increase sharply. Adding fillers with low activity may be effective to replace part of cement and sand by reducing the thickness of the glue layer on the aggregate grains and decreasing their voidness. The optimum amount of dispersed fillers in cement systems varies in the range between 10 and 70%.

There are conflicting recommendations for optimum dispersion of fillers. For example, it is recommended to use more fine clinker component when adding inert or low-active fillers in cement. However, it was shown recently that the increased strength of dispersed systems (in the presence of superplasticizers) is due to the use of fillers with a dispersion of 900 ... 1200 m^2 / kg, that is, 2–4 times higher than cement dispersion. Theoretical studies showed that for cement-water systems filled with fine quartz sand, the optimal ratio between the size of filler and binder should be within the range of 8 ... 10. However, if the surface energy of the filler granules is substantially higher than that of the cement grains, the desired ratio of their particles diameters is approaching 1.

Predicting the influence of fillers on the cement composites strength, it is necessary to take into account the chemical and mineralogical composition of clinker. For example, active fillers such as fly ash, increase the strength at early hardening stage due to the high content of alkalies in the clinker, which accelerates the chemical interaction of filler and cement. When using all kinds of fillers with pozzolan activity, preference is given to cements with high alite content, which at hydrolysis forms Ca (OH)$_2$. Carbonate fillers are also chemically active in relation to hydroaluminates and hydrosulfoalluminates formed during cement hydration.

Along with the influence of active fillers on concrete and mortar compressive strength [43], they have a positive effect on the ratio of tensile strength to compressive strength and, accordingly on cement composites crack resistance. It is known that a higher ratio of concrete tensile to compressive strength occurs due to adding ash and slag in cement as well as by using carbonate fillers.

Depending on the type of mineral filler, its physical-chemical and chemical activity, the construction and technical properties of cement composites can change. Experimental results show that a possible increase in cement hydration degree by active fillers and concrete mixtures density, a decrease in concrete creep can be obtained.

The influence of fillers on shrinkage deformations is ambiguous. The cement stone shrinkage is determined by the amount of evaporated adsorption-coupled water and depends on the relative content of calcium hydrosilicates and their surface area. Therefore, adding finely dispersed high-active fillers enables to expect some increase in shrinkage deformations. At the same time, with reduced water demand it is possible to decrease the shrinkage deformations of filled cement composites. This is especially true when using fillers with a low chemical activity, which, as well as aggregates, reduces relative content of calcium hydrosilicates in the hardening system.

In addition to the strength and deformation properties, adding certain mineral fillers to the cement systems can be controlled, if necessary, corrosion resistance, exotherm, heat resistance, water resistance, etc.

However, along with the positive effect, adding mineral fillers can cause some negative consequences. For example, fillers containing soluble alkalies can cause an increase in reactivity of binder in concrete and mortars to aggregates with active silica. In this case, when using reactive fillers, the content of alkaline oxides in the filled binder should not exceed 0.6% by weight and it is desirable to use practically alkaline-free cements. At the same time, some studies have shown that replacing cement with filler, such as coal ash reduces the interaction between alkalies and aggregates and the permissible limit of the possible alkali oxides content in the cement-ash binder can be raised. Special experimental studies are required when adding mineral fillers into cement composites with high frost resistance, abrasion resistance, hardening at low temperatures, and others.

2.3 Ways of activating mineral fillers

As already noted, activity of mineral fillers is obtained due to their-physical-chemical and chemical influence on the artificial stone formation processes [28, 44, 45]. Increase in the fillers activity (activation) can be achieved by a set of technological methods, aimed at increasing their positive effect in cement systems. Classification of fillers activating methods with indication of their possible quantity as a percentage of the cement mass and the expected strength growth is given in Table 2.1.

Table 2.1 Methods of activating fillers and their influence on concrete strength.

Activation method	Filler content, relative to the mass of cement, %	Increase in concrete strength, %
- milling with creation of protective films	5 ... 15	125 ... 140
- using active additives to increase surface energy	15 ... 30	130 ... 155
- heat treatment to glass state	3 ... 18	150
- hydro-thermal processing	30 ... 50	145 ... 165
- hydro-mechanical processing with lime-containing components	30 ... 40	150 ... 170
- treatment by strong acids solutions	10 ... 25	150 ... 165
- treatment by hydrophilizing SAS	10 ... 25	115 ... 120
- acoustic processing	≤ 100	140 ... 180
- electrodynamic processing	≤ 100	150 ... 170

Increasing the specific surface of the filler by additional grinding (mechanical method) and modifying its surface by addition of SAS (chemical method) are the main activation methods that were investigated. Combination of both methods (mechanical-chemical method) is also effective.

At additional milling of fillers, their chemical potential μ increases as follows:

$$\Delta\mu = 2\sigma V_m / r \qquad (2.5)$$

where σ is the surface energy value; V_m is the quantity (in mole) of the substance forming a spherical particle; r is the particles radius.

The crystalline structure of minerals, partial amorphization, and formation of unsaturated valence bonds have a significant influence on increasing the powder activity by grinding. The desired ratio between the size of filler particles d_f and binder grains is developed based of theoretical concepts of clusters formation and the self-organization of the cement stone structure. For the most common cases when the binder surface energy (σ_b) is higher than that of the filler (σ_f) it is recommended according to these concepts that $d_f/d_b = 3 \ldots 10$. At $E_f \geq E_b$ is recommended $d_f/d_b \approx 1$. In these cases at excessive reduction in the filler grains size the probability of their unification in cluster structure increases, which leads to the danger of cracks formation.

To identify the surface energy of Portland cement and some types of fillers, their electronic spectra were studied, which allowed finding the concentration of acid and basic centers on the surface of these powders. As a surface activity criterion the ratio of the total concentration of acidic and basic centers is taken. It is 8.8, for Portland cement, 16.4 for slag and 15.4 for fly ash. However such a way of determining the powders surface activity is rather complicated. In practice indirect methods are used by the strength of the particles adhesion and the wetting heat in various fluids.

The possibility of mineral fillers activation by adding SAS follows from the Dupree-Young equation, which relates the work of adhesion W_{ad} to the surface energy W_S:

$$W_{ad} = W_S - W_S \left(m + \cos\theta \right), \qquad (2.6)$$

where W_S is the free surface energy of a solid body in vapor and gas atmosphere; $m = \sigma_l' / \sigma_l > 1$ (σ_l' is the surface tension of liquid, oriented under the influence of the force field of a solid surface; σ_l is the liquid surface tension); θ is the edge wetting angle.

When processing the filler with SAS additives as a result of the adsorption medium creation, there is a decrease in the phase surface energy $\Delta_{.m.l}$ and, consequently, an increase in the adhesion work:

$$\Delta\sigma_{m.l.} = KT \int_0^c n_s(C)d\ln C \qquad (2.7)$$

were K is the constant of Boltzmann; T is the absolute temperature, $^{\circ}K$; n_s is the value of adsorption, caused by SAS molecules, adsorbed at 1 cm^2 of the phase separation surface (S); C is the SAS concentration.

The influence of the adsorption medium created by SAS, increases proportionally to the filler dispersion and, accordingly, to the excess surface energy. Activation of the mineral powders adhesive ability can be achieved by increasing their free surface energy under the action of electric and magnetic fields, ultrasonic treatment, or by means of ionizing radiation.

For producing mortars and concrete there are more interesting ways of activating fillers that are easy to implement and require minimal energy consumption.

2.4 Scientific hypothesis

Theoretical analysis shows that the dispersed fraction, which is part of stone crushing waste like siftings and aspiration dust, under certain conditions can have a positive effect on the structure of cement systems. At the same time, its integral influence on the properties of concrete and mortars can be positive or negative. The negative effect of the dispersed mineral fraction, which is a filler, on the properties of concrete and mortars may be mainly due to a significant increase in water demand to provide the required mixtures workability and, as a consequence, an increase in the total concrete and mortars porosity.

The scientific hypothesis that is experimentally verified by the authors and discussed in the book, is that in order to neutralize the negative and to obtain the positive effect of the dispersed fraction of stone crushing waste added into cement systems, it is necessary to prevent a significant increase in water demand of mixtures. This is possible in two ways: (1) adding effective plasticizers into concrete or mortar mixtures; (2) using stiff mixtures compacted by compression or vibration. To realize these technological solutions it is important to ensure the optimal grain composition of waste and the content of dispersed fractions in them, taking into account the purpose and composition of concrete or mortar mixtures.

Plasticizers, according to their efficiency, i.e., reducing action or increasing the mixtures workability without decreasing the concrete strength, are divided into 4 categories (Table 2.2). The most effective plasticizers—superplasticizers began to be used in concrete and mortar production in the early 1970s. These plasticizers allowed significant improvement in properties of concrete and mortar without increasing the content of cement, applying cast and self-compacting mixtures with moderate water demand.

Addition of superplasticizers to reduce the water-cement ratio is a prerequisite for producing high performance concrete (HPC).

A common classification of superplasticizers (Table 2.3) divides them by composition and action caused by a complex of physical and chemical processes in the cement paste—additive system:

1) SAS adsorption on the surface of mainly hydrated compounds;
2) colloid-chemical phenomena on the phase separation border.

Table 2.2 Classification of plasticizers for concrete mixtures.

Category	Name	Effective plasticizing action (increase of slump from 2 ... 4 cm), cm	Water demand reduction, %
I	superplasticizers	20 and more	at least 20
II	plasticizers	14 ... 19	at least 10
III	plasticizers	9 ... 13	at least 5
IV	plasticizers	≤ 8	< 5

Table 2.3 Classification of superplasticizers (SP).

Marking	SP content	Mechanism
NF	Based on sulfated naphthalene formaldehyde polycondensates	electrostatic
MF	Based on sulfated melamin-formaldehyde polycondensates	electrostatic
LST	Based on lignosulfonates purified from sugars	electrostatic
P	Based on polycarboxylates and polyacrylates	steric

In the action of superplasticizer types NF, MF, LST, Table 2.3 prevails the effect of electrostatic repulsion of cement particles, due to the fact that the adsorption layers of the superplasticizer molecules increase the value of the zeta potential on the cement particles surface. The zeta potential magnitude with a negative sign depends on the adsorption capacity of the superplasticizer. The superplasticizer adsorption capacity increases if the hydrocarbon chain length and the molecular mass increase. The superplasticizer adsorption is proportional to its concentration in aqueous solution.

Of cement minerals, the highest adsorption capacity have tricalcium aluminate (C_3A), and the lowest - β-two-calcium silicate $(\beta\text{-}C_2S)$.

In the action of last generation type P superplasticizers the role of the zeta potential is lower and mutual repulsion of the cement particles is provided by the so-called steric effect. This effect is due to the chains shape and the nature of the charges on cement and hydrates grains surface. Additives on the polycarboxylates basis provide an increase in the concrete mixture cone slump from 3 to 21 ... 24 cm with a dosage of only 0.17 ... 0.22% of the cement mass. If concrete mixtures with additives of traditional superplasticizers quickly lose workability, those with polycarboxylates remain in plastic state 1.5 ... 2 hours. High capacity for storage of concrete mixtures with type P superplasticizers makes them especially attractive for monolithic construction and at long transportation time. Like other superplasticizers, they are successfully applied in heat-treated concrete technologies in the industry of prefabricated reinforced concrete. High-reducing ability of polycarboxylate superplasticizers (up to 30%) makes them particularly attractive for mixtures containing highly dispersed fillers.

When obtaining concrete with stone crushing waste as a filler, using stiff concrete mixtures should be considered that along with some increase in water demand of such mixtures with increase in the dispersed fraction content, concrete forming and density at force compacting improve. As a result, appropriate conditions are created for compensation of negative and formation of positive effect of highly dispersed fillers.

CHAPTER 3

Fine-Grained Concrete Based on Stone Siftings

3.1 General characteristics

The maximum size of the aggregate in fine-grained concrete (FGC) is limited [46] to 10 mm. A common type of this concrete is sand concrete that does not contain coarse aggregates. From the viewpoint of using siftings of granite or other rocks, fine-grained concrete is the most promising material, since it enables to dispose of more material and to fully realize its advantages.

High specific surface area of aggregates in fine-grained concrete under traditional technology causes an increase by 20 ... 40% in cement consumption, which is required to fill the intergranular cavities and to create a sufficient layer of cementitious coating on the aggregate grains. Reducing cement consumption is achieved by choosing the optimum granulometric composition of the aggregate, adding active mineral fillers, use of superplasticizers and effective compacting methods [47, 48].

When using fly ash as filler, the degree of filling (F/C) is 0.4 ... 0.6, for microsilica it is 0.06...0.15. Here F and C are the quantities of active additive and cement respectively. The optimum degree of filling of fine-grained concrete by dispersed mineral additives depends on the water demand of their particles, the chemical activity in relation to $Ca(OH)_2$ and participation in the cement stone structure formation processes.

Fine-grained concrete is characterized by an increased ratio between tensile strength and compressive strength (Fig. 3.1). At the same compressive strength, tensile strength for fine-grained concrete is 10 ... 15% higher when compared to the conventional one. Accordingly, the concrete dynamic strength and endurance increase. This is due to the higher homogeneity of a fine-grained concrete structure.

The FGC structure features also affect its deformation properties. They have a modulus of elasticity of 20 ... 30% lower than conventional concrete, with higher shrinkage and creep values. Deformability and creep can be significantly reduced, due to increased stiffness of concrete mixtures, and the use of power compacting methods.

$f_{c.tk}, f_{c.tf}$, MPa

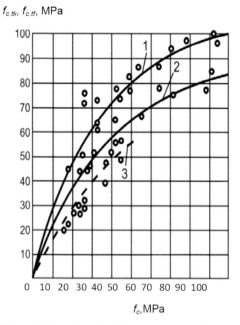

f_c, MPa

Fig. 3.1 Dependence of flexural ($f_{c.tf}$) and tensile ($f_{c.tk}$) concrete strength on compressive strength (f_c): $1 - f_{c.tf}$ FGC; $2 - f_{c.tf}$ conventional concrete; $3 - f_{c.tk}$ FGC.

The FGC mixtures workability is measured in three ways: by a standard cone slump, depth of cone penetration and the cone spread on the shaking table. Indicators, measured by these methods, are interdependent (Table 3.1). The cement-sand mixture, having the same cone slump as the conventional concrete mixture, is compacted better and faster [49].

Different technologies for producing FGC with improved properties have been developed. These technologies include additional cement milling, joint milling of cement and sand, using vibrating and jet mixers, applying intensive compacting methods such as vibration stamping, vibration pressing, semi-dry pressing, roller forming, etc.

Table 3.1 Workability of fine grained concrete mixtures, obtained by different methods.

Standard cone slump (CS), cm	Cone penetration depth, cm	Cone flow on the shaking table, mm
1 ... 3	2 ... 3	110 ... 140
3 ... 6	3 ... 5	140 ... 170
5 ... 8	4 ... 6	160 ... 180
8 ... 14	6 ... 8	170 ... 200
12 ... 15	7 ... 9	190 ... 220
15 ... 22	8 ... 11	210 ... 240
20 ... 25	10 ... 14	230 ... 270

3.2 Influence of granite filler on fine-grained concrete properties

To investigate the influence of technological parameters on properties of concrete with the waste of rock crushing, the authors used the methods of mathematical experiments planning and experimental-statistical models based on these methods [50–53]. In their experimental research, mathematical experiment planning (MEP) is widely used. Following the MEP methodology, the experiments were carried out according to typical plans, which are optimal in terms of the experimental work and statistical requirements.

The experiment planning involves selecting the most significant factors and their variation limits to determine the desirable parameters, as well as conducting experiments on a certain statistically optimal plan (matrix of planning), the type of which is determined by the predicted dependence. The factors are presented in code and in natural values. Three-level plans were used to study the dependencies obtained in a wide range of factor changes.

For concrete mixtures, composition design it is possible to use two types of polynomials:

$$y = b_0 + \sum_{i=1}^{k} b_i \cdot x_i + \sum_{i=1}^{k} b_{ii} \cdot x_i^2 + \sum_{i \neq j}^{k} b_{ij} \cdot x_i \cdot x_j + ... \tag{3.1}$$

$$y = a_0 + \sum_{i=1}^{k} a_i \cdot V_i + \sum_{i=1}^{k} a_{ii} \cdot V_i^2 + \sum_{i \neq j}^{k} a_{ij} \cdot V_i \cdot V_j + ... \tag{3.2}$$

where b and *a* are statistical estimations of actual regression coefficients, *x* and V is investigated factors variables.

The difference between two polynomials is that if in the polynomial (3.1) factors x_i is independent, in polynomial (3.2) the following condition is satisfied:

$$\sum_{1}^{n} \varphi_i = 1 \tag{3.3}$$

where φ_i is the partial share of the *i*-th component in the concrete mixture; n is the number of components.

Polynomial models (3.2) are used when the mixture composition is given by K–1 specific components consumption or their ratios. The content of the component φ_k is not varied according to the planning matrix and is found from the material balance condition:

$$\varphi_k = 1 - \sum_{i=1}^{k-1} \varphi_i \tag{3.4}$$

In concrete composition design tasks, polynomial regression equations can be used as ordinary quantitative dependencies for certain boundary conditions. If quadratic polynomial models with $x_1, x_2 \ldots x_n \ldots x_k$ given factors are obtained for certain parameters, then setting the value of a particular factor X_n, for example, cement-water ratio, it is possible when presenting the models in the form of square equations:

$$b_0 + \sum_{i=1}^{k} b_i \cdot x_i + \sum_{i=1}^{k} b_{ii} \cdot x_i^2 + \sum_{i \neq j}^{k} b_{ij} \cdot x_i \cdot x_j - y_i = 0 \tag{3.5}$$

By setting the certain value of the parameter (y_i) and stabilizing other factors at a certain level, it is possible to find the value of x_n as the root of the square equation:

$$x_n = \frac{-C_n \pm \sqrt{C_n^2 - 4b_{nn}l}}{2b_{nn}} \tag{3.6}$$

where $C_n = b_n + \sum_{\substack{i=1 \\ i \neq n}}^{k} b_{ni} x_i,$

$$l = \sum_{\substack{i=1 \\ i \neq n}}^{k} b_i x_i + \sum_{\substack{i=1 \\ i \neq n}}^{k} b_{ii} x_i^2 + \sum_{\substack{i=1 \\ i \neq j}}^{k} b_{ij} x_i x_j - y_i.$$

Transition to the value of the factor in physical units is achieved using the transformation formula:

$$x_i = \frac{\tilde{x}_i - \tilde{x}_{i0}}{\Delta \tilde{x}_i} \tag{3.7}$$

where x_i is the coded value of the factor; $\tilde{x}_i$ is value of the factor in physical units; $\tilde{x}_{i0}$ is the factor value of in physical units on main level; $\Delta \tilde{x}_i$ is the factor variation interval in physical units.

Polynomial (3.2) is usually presented as follows:

$$y = \sum A_i \cdot V_i + \sum A_{ij} \cdot V_i \cdot V_j \tag{3.8}$$

where $A_i = a_0 + a_i + a_{ii}$ $A_{ij} = a_i + a_{ii} + a_{ij}$.

Simultaneous variation of the mixture K-components and obtaining of the polynomial (3.1) is possible with the use of partial relations:

$$x_1 = \frac{\varphi_1}{\varphi_1 + \varphi_2}; \ldots x_{k-1} = \frac{\varphi_1 + \varphi_2 + \ldots + \varphi_{k-1}}{\varphi_1 + \varphi_2 + \ldots + \varphi_k} \tag{3.9}$$

where φ is the volume component concentration.

By simple perturbations it is possible to find volumetric parts of each component in the mixture:

$$\varphi_1 = x_1 x_2; \quad \varphi_2 = (1 - x_1) x_2; \text{ etc.}$$

At planning relations (3.9) a polystructural approach to investigating the composition of mixtures and materials on their basis is realized. The structural level of factors is successively changed as the number of components increases.

To obtain polynomial models of type 3.1, different standard plans are used to vary the investigated factors at two, three or more levels, optimizing the number of experiments and other statistical parameters.

To obtain polynomial models of type 3.2 plans are used that allow optimal location of experimental points on a simplex-shape, formed by a set of (K + 1) independent points in a K-dimensional space with minimum number of vertices (triangle, tetrahedron, etc.). The most well-known of the simplest plans is the Sheffe plans, which have uniform location on the experimental points' simplex (simplex grades). Polynomial models such as 3.2 directly allow constructing and investigating the property-composition diagram on a simplex or its plane projections. In Sheffe plans the number of experimental points is minimized: so when K = 3 it is 6 and 10 in the plans of the second and third order, respectively.

Since the experiment is necessarily associated with errors, because the obtained results are probabilistic, and the adopted regression equations are not their copy, but only reflect them with a certain degree of probability, therefore a compulsory stage during the research is statistical analysis of mathematical models. The main purpose of this analysis is to evaluate the significance of the coefficients in the equations and verify their adequacy. Statistical analysis of mathematical models was carried out on the basis of repeated experimental data at the main (zero) level.

The mean value of the studied parameters were obtained as follows:

$$\overline{y} = \frac{\sum_{i=1}^{n} y_{oi}}{n} = \frac{y_{o1} + y_{o2} + \ldots + y_{on}}{n} \tag{3.10}$$

where y_{oi} is the value of the parameter on the zero level; n is the number of repeated zero points.

Dispersion of the studied (output) parameter reproducibility S_r^2 was calculated as:

$$S_r^2 = \frac{\sum_{i=1}^{n} (y_{oi} - \overline{y}_o)^2}{n - 1} \tag{3.11}$$

The number of degrees of freedom of the reproducibility dispersion is $f_r = n - 1$. The error of the output parameter values was determined by the formula:

$$\Delta y_u = \pm \frac{S_r}{\sqrt{n}} t_{1-p/2} \tag{3.12}$$

where t is quantile distribution of Student; p is confidence probability level.

Dispersion of the experimental points' adequacy relative to the mathematical model S_a^2 is obtained as:

$$S_a^2 = \frac{\sum\limits_{u=1}^{n} (y_u - \bar{y}_u)^2}{N - m - (n-1)} \tag{3.13}$$

where y_u is experimental value of the parameter in u-throw of the matrix; $\bar{y}_u$ is calculated value of the parameter in u-throw of the matrix by mathematical model; N is the number of rows in the matrix; m is the number of significant coefficients in the mathematical model.

The number of degrees of freedom of the dispersion reproducibility is

$f_a = N - m - (n-1)$.

The estimated value of Fisher's criterion is:

$$F_p = \frac{S_a^2}{S_r^2} \tag{3.14}$$

The table value of Fisher's criterion F_t was obtained depending on the accepted confidence probability and the number of freedom degrees. In concrete technology, the confidence probability is assumed to be equal 95% (90%). The equation is considered adequate if $F_p < F_t$.

The error of the output parameter values obtained by the mathematical model was determined as:

$$\Delta y_u = \pm \frac{S_a}{\sqrt{N}} t_{1-p/2} \tag{3.15}$$

Statistical analysis of the experimental results, calculation of regression equations coefficients and graphic dependencies, iso-parametric analysis of models as well as obtaining of nomograms was carried out using a package of special computer programs developed in the frame of this work. The programs are realized in Microsoft Excel environment.

To evaluate the combined effect of fillers and plasticizers on FGC properties, algorithmic experiments were carried out using as a filler fraction < 0.16 mm of granite siftings and for comparison fly ash with naphthalene-formaldehyde superplasticizer. A typical B_4 plan of experiments was used.

As main factors that affect FGC properties were accepted: consumption of naphthalene—formaldehyde type superplasticizer, degree of filling of cement F/C, cement consumption, and cement-water ratio C/W (Table 3.2). The following parameters were studied: compressive strength at 28 days, MPa and cone flow on shaking table (CF, mm).

Table 3.2 Experiment planning conditions for obtaining Eqs. 3.1–3.4

Factors		Factors variation levels			Variation interval
Natural	**Coded**	**−1**	**0**	**+1**	
Superplasticizer content (SP), %	X_1	0	0.5	1	0.5
Degree of filling F/C: – particles < 0.16 mm of granite siftings; – fly ash	X_2	0	0.45	0.9	0.45
Cement consumption (C), kg/m³	X_3	340	400	460	60
C/W	X_4	1.68	1.89	2.30	0.41

As a result of the statistical processing of experimental data, regression equations were obtained which can be considered as mathematical models of concrete compression strength (f_c) and concrete mixture cone flow (CF) for two types of fillers:

Filler – granite siftings particles < 0.16 mm:

$$f_{c,s}^{28} = 29.6 + 2,21x_2 + 6.54x_3 + 1.17x_4 - 2.2x_1^2 -$$
$$-1.7x_2^2 - 2.45x_3^2 - 2.65x_4^2 + 1.66x_1x_2 - 1.39x_1x_3 +$$
$$+1.57x_2x_3 + 3.02x_3x_4 \tag{3.16}$$

$$CF_s = 129.7 + 12.1x_1 - 5.1x_2 + 20.8x_3 - 19.8x_4 +$$
$$2.1x_1^2 - 2.9x_2^2 + 3.1x_3^2 - 3.9x_4^2 + 5.8x_1x_3 -$$
$$4.7x_1x_4 - 3.2x_2x_3 - 13.4x_3x_4 \tag{3.17}$$

Filler – fly ash:

$$f_{c,a}^{28} = 34.15 + 5.1x_2 + 5.07x_3 + 0.26x_1^2 - 5.54x_2^2 -$$
$$4.34x_3^2 - 1.04x_4^2 + 1.51x_1x_2 - 1.57x_1x_3 +$$
$$1.24x_1x_4 - 1.16x_2x_4 + 3.02x_3x_4 \tag{3.18}$$

$$CF_a = 161.4 + 18.5x_1 + 9.2x_2 + 24x_3 - 27.2x_4 -$$
$$5.6x_1^2 - -24.1x_2^2 - 7.1x_3^2 + 16.4x_4^2 + 9.6x_1x_2 +$$
$$7.7x_1x_3 - 6.1x_1x_4 - 6.4x_2x_4 - 11.1x_3x_4 \tag{3.19}$$

Analysis of the regression coefficients in Eqs. 3.16 and 3.18 shows that the dependence of concrete strength on the investigated factors is nonlinear. Depending on the impact degree on the compressive strength by absolute values, the factors in the selected variation intervals are arranged in the following sequence:

$$SP- C>F/C>C/W$$

The strongest interactions of the factors are between cement consumption and cement-water ratio, degree of filling and cement consumption. Interaction between consumption of superplasticizer and cement consumption is also noticeable.

The quadratic effects in the Eqs. 3.16 and 3.18, with a negative sign indicates that increasing the value of each factor from a minimum of -1 to a maximum of $+1$ at other equal conditions starting from some point causes a decrease in strength, i.e., a zone of optimum strength in the investigated factor space is evident.

Of particular interest is the nature of FGC strength dependence on the degree of filling with granite siftings particles < 0.16 mm.

Differentiation by X_2 enables to obtain the optimum equation:

$$x_{2opt} = 1/3.4\left(2.21+1.68x_1 +1.57x_3\right) \tag{3.20}$$

The optimum degree of filling $(F/C)_{opt}$ by granite siftings fraction < 0.16 mm at adding naphthalene – formaldehyde type superplasticizer depends only on cement consumption and decreases with its growth.

Two-factor diagrams are constructed for technological interpretation of multi-factor regression equations. They were obtained by solving regression equations relative to two variables when fixing other factors at a certain level (Figs. 3.2, 3.3). At each intersection a set of uniform strength curves is shown, which makes it possible to predict the influence of factors X_2 and X_3 when fixing others at the main levels. Isolines characterize the possible ratio of factors that provide the same strength.

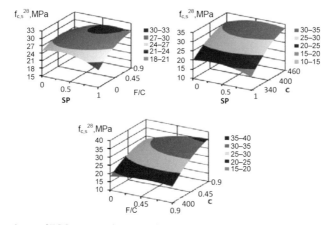

Fig. 3.2 Dependence of FGC compressive strength at 28 days when using dust fraction of granite siftings on the investigated factors.

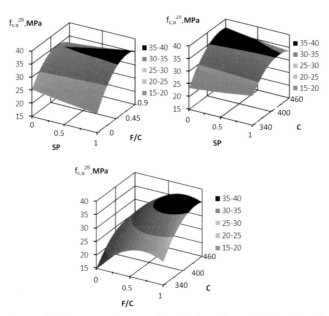

Fig. 3.3 Dependence of FGC compressive strength at 28 days when using fly ash on the investigated factors.

Calculation of the optimal superplasticizer consumption shows that for superplasticizer consumption $X_{1opt} = 0$, i.e, 0.5%, as it follows from Fig. 3.2, a zone of optimum strength at $f_{c,s} > 30$ MPa is evident. It is characterized by the following FGC composition parameters:

$$C = 460 ... 444 \text{ kg/m}^3; \text{ F/C} = 0.375 ... 0.5; \text{ C/W} = 2.19 ... 2.3$$

In order to evaluate the joint impact of technological factors, the strengths graphics were also plotted depending on the degree of filling F/C (X_2) and consumption of additive of superplasticizer (X_1) with fixed values cement content (X_3) and cement-water ratio (X_4) on the main (zero) level (Fig. 3.2). When using of granite siftings particles < 0.16 mm at C/W = 1.68 in the established factors space there is a zone of optimal consumption of filler and additive. At C/W = 1.68 and cement consumption of 336 ... 460 kg/m³ the optimum degree of filling increases with the increase of superplasticizer content. In concrete without superplasticizer (F/C)$_{opt}$ = 0.25 ... 0.35. If the superplasticizer content is 0.5% by cement mass (F/C)$_{opt}$ = 0.35 ... 0.45 if the superplasticizer content is 0.5 ... 1% (F/C) = 0.45 ... 0.55.

At C/W = 2.3 (Fig. 3.2) and the cement consumption of 336 ... 460 kg/m³, the optimum degree of filling does not increase with the increase of the superplasticizer consumption, but depends only on cement consumption. The superplasticizer content in these conditions has an optimal value of 0.45 ... 0.65% by the cement weight. This is explained by the fact that at C/W = 2.3 the required water demand for the concrete mixture workability is not provided. In such conditions, plasticizers just eliminate the negative effects of fillers on the concrete mixture rheological parameters.

Using granite siftings particles < 0.16 mm and superplasticizer enables to increase the concrete strength by 10 ... 20% in comparison with filled FGC without superplasticizer and by 20 ... 30% in comparison with plasticized concrete without fillers. The nature of the fly ash impact on FGC compressive strength according to the Eq. 3.3 is given in Fig. 3.3. Its analysis shows that an increase in cement consumption in the variations range can in different ways affect the concrete strength at constant C/W.

Positive influence of cement consumption decreases with its increase in the concrete mixture. In some areas at C/W ≤ 1.99 and considerable content of SP admixture there is a negative effect [54, 55]. Data on strength reduction when increasing the cement consumption under constant C/W is also confirmed by other researchers [56]. An increase in cement consumption over 400 kg/m³ with a high degree of filling by ash and a constant C/W causes an increase in water demand and, consequently, capillary porosity.

When forming the concrete structure, the strength between the cement stone contact layers and the aggregate is important. Following the modern concepts of concrete physical and chemical mechanics, the cement stone strength in contact layers on the border with quartz sand grains is 1.5 times and higher than that of the stone in volume. In this case, the influence of contact layers on concrete strength increases with the distance between the aggregate grains decreases, that is, the strength depends on the ratio between the contact layer size and distance between the grains, as well as the fine filler volume concentration. Hence an increase in the cement paste quantity at a certain point begins to play a negative role on the structure and strength of concrete.

Finding optimal parameters of FGC filled by ash, shows that concrete compressive strength of 35 ... 40 MPa can be achieved for a wide filling degree range. At the same time, the minimum F/C value corresponds to the maximum C/W and cement consumption over 430 kg/m³, and the maximum value corresponds to C/W = 1.99 ... 2.3 and average cement consumption of 400 ... 430 kg/m³.

When adding to the concrete mixture fly ash, the content of superplasticizer (X_{1opt}) depends on the cement-water ratio. The higher the value of C/W, the lower the X_{1opt}. This can be explained by the fact that fly ashes have a plasticization effect in some optimal filling range [57].

The optimal parameters of FGC filling by ash and plasticization by superplasticizer are clearly expressed at high C/W values, for which $(F/C)_{opt.} = 0.6 ... 0.7$, and the optimal superplasticizers content is within 0.5 ... 0.6% by cement weight. The highest relative effect of filling and plasticization is achieved at relatively small cement consumption (C = 340 kg/m³).

From Eqs. 3.2 and 3.4, it can be noted that as the cement paste content and the superplasticizer content increase, the mixture workability leads to a decrease in the cone flow. With the increase in the admixtures consumption and filling degree, the effect of the latter factor on the concrete mixture workability parameters is compensated. Figure 3.4 presents an iso-parametric diagram for workability of mixtures containing particles of granite siftings < 0.16 mm and naphthalene-formaldehyde superplasticizer. As can be seen from the diagram the superplasticizer efficiency decreases with increasing the filling degree.

The highest decrease in concrete mixtures workability, as shown by analysis of Eqs. 3.2 and 3.4, occurs when the cement and filler consumptions increase at simultaneous increase of C/W. Because of the negative sign of most quadratic effects in Eqs. 3.2 and 3.4, it can be concluded that there exists a zone of optimum cone flow for FGC with filler.

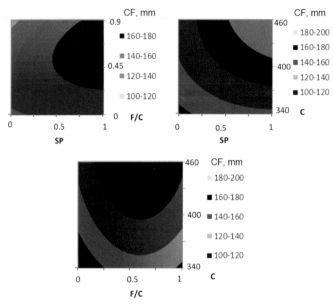

Fig. 3.4 Dependence of concrete mixtures workability with SP filled by fly ash on the investigated factors.

When applying superplasticizer, the optimum filling degree can be assumed to be 0.5 ... 0.65. This optimum is valid for different C/W values. With an increase in cement consumption there is a tendency to decrease the optimum by 0.05 ... 0.1, which is confirmed by available data [18].

Analysis of the obtained dependences enables to suggest that joint addition of fillers (particles of granite siftings < 0.16 mm or ash) and superplasticizers into the concrete mixture in all cases is an effective method to increase concrete strength at a given concrete mixture workability and reducing the FGC cement consumption.

3.3 Influence of granite filler dispersion on FGC properties

To study the influence of granite siftings particles < 0.16 mm dispersion of on the properties of FGC, a planned experiment was conducted for four factors according to B_4 plan. As varied factors are taken:

X_1 – cement consumption, kg/m³ of concrete mixture;

X_2 – superplasticizer content (SP), % of cement mass;

X_3 – content of particles < 0.16 mm in siftings (m_{016}, %);

X_4 – specific surface of particles < 0.16 mm S_{016}, cm²/g.

The increase in the surface of the particles < 0.16 mm was carried out by partial or complete replacement of the disperse particles contained in the siftings, by aspiration granite dust with a specific surface of 4700 cm²/g. The concrete mixture fluidity was controlled by measuring the cone flow on the shaking table and at all points of the plan was kept at a level of 180 ± 5 mm. Compacting was carried out by vibration. The experiments planning conditions are given in Table 3.3.

Table 3.3 Experiments planning conditions for Eqs. (3.6 … 3.9).

No.	Factors			Variation levels			Variation interval
	Natural	**Coded**		−1	0	1	
1	Cement consumption, kg/m³	X_1		200	350	500	150
2	SP content, % of cement weight	X_2		0	0.6	1.2	0.6
3	Content of disperse particles in sifting, ($m_{0.16}$, %)	X_3		0	12	24	12
4	Specific surface of particles < 0.16 mm, $S_{0.16}$, cm²/g	X_4		2300	3500	4700	1200

Experimental-statistical models of output parameters (water-cement ratio (W/C), average density of FGC (ρ_0, g/cm³), compressive strength (f_c^{28}, MPa) tensile strength in bending ($f_{c.tf}^{28}$, MPa) at 28 days are:

$$W/C = 0.62 - 0.28x_1 - 0.07x_2 + 0.17x_3 + 0.09x_4 +$$
$$0.11x_1^2 + 0.02x_2^2 - 0.04x_3^2 + 0.04x_4^2 + 0.005x_1x_2 - \qquad (3.21)$$
$$0.11x_1x_3 - 0.08x_1x_4 - 0.05x_2x_3 - 0.01x_2x_4 - 0.06x_3x_4$$

$$\rho_0 = 2.12 + 0.09x_1 + 0.01x_2 + 0.07x_3 + 0.01x_4 +$$
$$0.13x_1^2 - 0.04x_2^2 - 0.06x_3^2 - 0.02x_4^2 - 0.01x_1x_2 - \qquad (3.22)$$
$$0.03x_1x_3 - 0.01x_1x_4 + 0.05x_2x_3 - 0.02x_2x_4 + 0.02x_3x_4$$

$$f_c^{28} = 33.96 + 9.94x_1 + 2.71x_2 + 2.78x_3 - 4.4x_1^2 -$$
$$2.15x_2^2 - -5.06x_3^2 - 4.61x_4^2 + 1.59x_1x_2 - 1.4x_1x_3 - \qquad (3.23)$$
$$0.82x_1x_4 + 1.99x_2x_3 + 0.27x_2x_4 - 0.5x_3x_4$$

$$f_{c.tf}^{28} = 8.857 + 2.62x_1 + 0.99x_2 + 0.3x_3 - 0.11x_4 -$$
$$1.65x_1^2 - 0.39x_2^2 - 0.544x_3^2 + 0.52x_1x_2 + \qquad (3.24)$$
$$0.19x_1x_3 - 0.18x_1x_4 + 0.8x_2x_3 - 0.14x_3x_4$$

The water demand of FGC mixture was evaluated for W/C value sup the cone flow of 180 mm, corresponding to CS = 9 ... 15 cm. The W/C was within the range of 0.4 ... 1.4. The water demand ranged from 100 to 330 l/m^3. The graphs of W/C dependence on the investigated factors are presented in Figs. 3.5 ... 3.12.

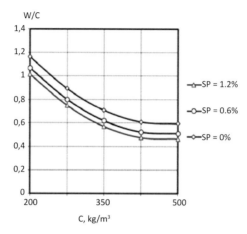

Fig. 3.5 Dependence of W/C on cement and superplasticizer content.

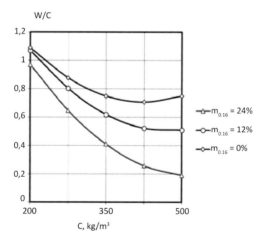

Fig. 3.6 Dependence of W/C on cement consumption and disperse particles content in siftings.

Analyzing the influence of factors, it should be noted that the highest linear effect on W/C has factor X_1 (cement consumption).The influence of factor X_3 (content of particles < 0.16 mm) is close to that of X_1. If the change of factor X_1 at transition from the lower to the upper variation level causes a decrease in W/C, then factor X_3, as expected, significantly increases the concrete mixture water demand. In turn, an additional increase in water demand also causes an increase in factor X_4 (specific surface of granite filler).

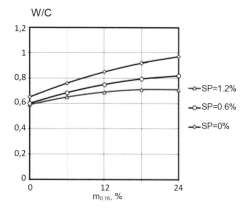

Fig. 3.7 Dependence of W/C on disperse particles and superplasticizer content.

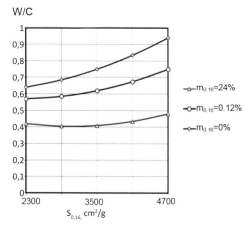

Fig. 3.8 Dependence of W/C on disperse particles content in siftings and their specific surface.

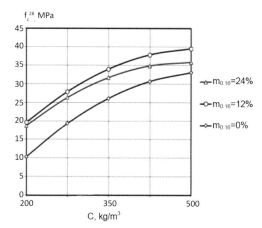

Fig. 3.9 Influence of cement consumption and disperse particles content on FGC compressive strength at 28 days.

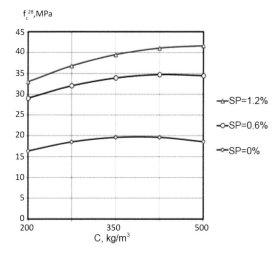

Fig. 3.10 Influence of cement consumption and superplasticizers content on FGC compressive strength at 28 days.

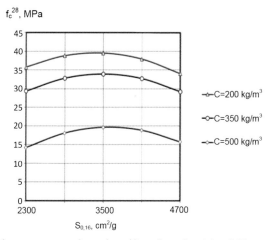

Fig. 3.11 Influence of cement consumption and specific surface of particles<0.16 mm on FGC compressive strength at 28 days.

Factor X_2 (superplasticizer content) to some extent compensates for the negative effect of dispersed particles, but, as it can be seen from the absolute magnitudes of the corresponding effects, this compensation is not sufficient in this case. The most important factor interactions in this model are those that can reduce the concrete mixture water demand at constant cement consumption. Thus, significant negative effect of b_{23} indicates some compensation for an increase in water demand, caused by disperse particles content increase at the superplasticizer consumption. The interaction of factors X_2 and X_4 is much lower—the diluting effect of naphthalene-formaldehyde superplasticizer is not enough to cover the effect on the water demand of disperse granite with a specific surface of 4700 cm^2/g.

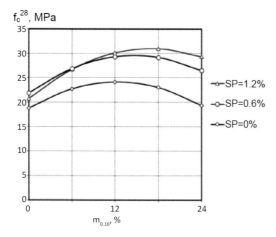

Fig. 3.12 Influence of superplasticizer sand disperse particles content on FGC compressive strength at 28 days.

Another indicator, characterizing the properties of concrete mixes with granite siftings is the average concrete mixture density after compacting. At constant cement consumption, the most influential factor, contributing to the concrete compacting, is the increase in the disperse particles content in granite siftings (X_3). Some compacting improvement is also observed due to increase of the superplasticizer admixture and granite dust dispersion (X_2, X_4) content. Significant negative quadratic effects indicate the extreme nature of the dependencies. An increase in the content of disperse particulate particles in sifting above 10 ... 12% reduces the concrete density, probably due to increased water demand of the mixtures.

Cement content (X_1) has the highest impact on strength. It causes a corresponding change in W/C. At constant cement consumption factors X_2 and X_3 (the content of superplasticizer and particles < 0.16 mm) contribute to the strength increase to the same extent. Model 3.23, has a significant positive coefficient of these factors' interaction. Increasing the content of disperse particles in the filler with simultaneous addition of superplasticizer yields higher strength. The presence of significant quadratic effects of these factors are in Eq. 3.23, and indicates the existence of these factors threshold values, after which the strength begins to decrease. This is especially noticeable for factor X_3 (content of dispersed particles in granite siftings). Increase in the specific surface of granite filler yields higher W/C. If the content of particles < 0.16 mm in sifting is up to 12%, their impact on the concrete compressive strength is positive. Further increase in the content of particles < 0.16 mm leads to a decrease in strength.

Increasing the siftings' dusty component dispersion causes some increase in the FGC strength (within the range of 15 ... 20%), despite the increase in W/C. The maximum specific surface that causes the increase in strength is within the limits of 3200 ... 3500 cm^2/g.

The main condition for the positive effect of disperse particles, which are most conducive to the cement stone structure, is the thinner action of the superplasticizer. This is confirmed by the presence of a positive effect of these factors' interaction in the model 3.23.

The positive effect of granite siftings particles < 0.16 mm on concrete strength is also related to their effect on average density. The increased voidness of the siftings sand fraction in the absence of dust particles causes significant concrete mix stratification, especially at low cement consumption and high mixture fluidity. With such a combination of factors, the minimum average density of concrete was about 1900 kg/m³ and compressive strength, respectively, 1 ... 3 MPa. For low-cement concrete, dusty granite filler increases the cement paste volume, homogeneity and connectivity of the mixture, providing a dense and homogeneous concrete structure. At such conditions the compressive strength increases to 10 ... 15 MPa. The maximum FGC strength at optimal combination of factors is 40 ... 42 MPa and in the absence of granite filler it is 30 ... 32 MPa.

The obtained results have confirmed the possibility of using granite siftings as the main aggregate of FGC and demonstrated a positive effect of particles < 0.16 mm on strength, provided that they neutralize their negative impact on water demand due to addition of superplasticizer.

3.4 Composition design of fine-grained concrete with granite siftings

To calculate the FGC compressive strength, the following expression is usually used:

$$f_c = AR_c(C/W - b) \qquad (3.25)$$

where R_c is cement activity, MPa; A and b are coefficients.

Yu.M. Bazhenov [31] has proposed the following empirical dependence of sand concrete strength:

$$f_c = AR_c\left(\frac{C}{W + V_{e.a.}} - 0.8\right) \qquad (3.26)$$

Where A = 0.8 for high quality materials, 0.75 and 0.65 for medium and low quality, respectively; $V_{e.a.}$ is the involved air volume.

It has been shown that for each FGC component there is an optimal value of W/C, providing the highest strength values (Fig. 3.13). At low values of W/C (W/C ≤ 0.4) the concrete strength varies linearly depending on the cement—sand ratio (C/S). With a decrease of C/S below the optimal values, the FGC mixtures workability is reduced (Fig. 3.14), with increasing C/S the amount of excess water in concrete is higher, which also leads to a decrease in strength.

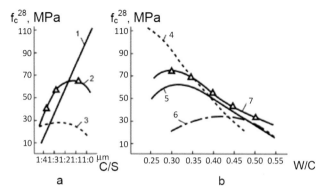

Fig. 3.13 Dependence of FGC compressive strength on composition (a) and W/C value (b): 1 – W/C = 0.3; 2 – W/C = 0.4; 3 – W/C = 0.5; 4 – C/S = 1.0; 5 – C/S = 1:2; 6 – C/S = 1:4; 7–normal-weight concrete (for comparison).

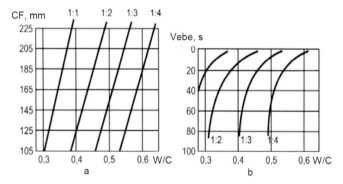

Fig. 3.14 Graphs for selecting the ratio between cement and medium fineness sand, providing the specified values of the (a) cone flow (CF) and (b) stiffness by the Vebe test of cement-sand mixtures (following Y.M. Bazhenov).

V.P. Sizov has proposed the following formula for calculating W/C of FGC:

$$W/C = \frac{(A + \sum \Delta A)R_c}{\dfrac{f_{cm}}{K_1 K_2} + 0.5(A + \sum \Delta A)R_c}$$

(3.27)

where $\sum \Delta A$ is a sum of corresponding corrections, depending on workability indicators, sand fineness and cement paste normal consistency. Coefficient K_1 depends on the cement mineralogical composition and varies from 0.88 for high aluminate ($C_3A > 10\%$) to 1.02 for high-belite cement. Coefficient K_2 depends on the concrete strength variation coefficient. At variation coefficient $V_c = 9 \ldots 12\%$ $K_2 = 0.98$, for $V_c = 13\%$ $K_2 = 0.95$ and for $V_c = 14 \ldots 17\%$ $K_2 = 0.92$.

The change of the dusty particles content in sand from 1 to 5% causes a decrease in the coefficient A from 0.52 to 0.47 [48]. Within these limits, respectively, the value of A decreases at transition from low flow-able to cast mixtures, decrease in the sand the sand fineness modulus from 3 to 1, and an increase in the cement paste normal consistency from 27% to 34%.

The filler quality affects the FGC basic properties to a greater extent than for ordinary normal-weight concretes. Replacement of coarse sand by a fine one can reduce the FGC strength by 25 ... 30%, and sometimes it becomes 2 ... 3 times lower, compared to that, obtained using coarse sand [31]. At optimal W/C and at equal concrete mixture workability using medium size aggregate is the most economical and provides the minimum ratio between cement consumption and concrete strength, achieved at ratio cement: sand (C/S) = 1:2 ... 1:3. When using fine-grained sand, 1:1 ... 1:1.5 are the optimal C/S ratios.

One of the FGC concrete mixtures features is an increased air retaining. The volume of the entrained air $V_{e.a}$ (l/m^3) can be found as follows:

$$V_{e.a} = 1000 - V_{c.m.}$$ (3.28)

where $V_{c.m.}$ is the volume of concrete mixture that changes, mainly with the changing concrete mixture stiffness.

The volume of entrained air depends on compacting method parameters and features.

Wide experimental data shows that the FGC compressive strength, in addition to C/W, cement activity and filler quality affects many other factors, such as concrete mixture workability, conditions of concrete hardening, presence and amount of active mineral admixtures, etc. For a given W/C, the ratio between the aggregate (sand) and cement (S/C = n) is uniquely determined by the mixture workability indicator (Fig. 3.15). Known values of W/C and n enable to find the content of all concrete mix components using the material balance equation (the sum of absolute volumes):

$$C = \frac{1 - V_{e.a}}{1/\rho_c + W/C + n/\rho_s}$$ (3.29)

$$W = C \cdot W / C$$ (3.30)

$$S = nC$$ (3.31)

Here ρ_c and ρ_s are the density of cement and sand, kg/m^3.

The content of aggregate in FGC mixtures can also be calculated based on understanding the mechanism of filling the volume of voids between the grains of sand and their extendable. In this case, the sand content can be found by solving the system of equations:

$$\frac{C}{\rho_C} + \frac{S}{\rho_S} + W + V_{e.a.} = 1$$ (3.32)

$$\frac{C}{\rho_C} + W = \alpha_e V_s^0 \frac{S}{\rho_{b.s.}} \tag{3.33}$$

where $\rho_{b.s.}$ is bulk density of sand, kg/m³; V_S^0 is the sand void ness; α_e is voids' filling and sand grains' extension coefficient.

Hence:

$$S = \frac{\rho_s \rho_{b.s.} (1 - V_{e.a.})}{\rho_{b.s.} + \alpha_e V_s^0 \rho_s} \tag{3.34}$$

Equation (3.34) can be re-written as:

$$S = \frac{\rho_{b.s.} (1 - V_{e.a.})}{1 + V_s^0 (\alpha_e - 1)} \tag{3.35}$$

According to experimental data [13], the coefficient α_e depends on the FGC mixtures workability indicators, C/W, sand fineness and cement paste normal consistency

Calculating the filler content according to Eq. 3.35 it is possible to find the cement consumption as a ratio of the cement paste volume in the concrete mixture ($V_{c.p.}$) to the cement paste obtained from 1 kg of cement ($V'_{c.p.}$).

$$V_{c.p.} = 1000 - V_s \tag{3.36}$$

where V_s is absolute volume of sand, *l* per 1 m³ of concrete mixture.

$$V'_{c.p} = \frac{1}{\rho_c} + W / C \tag{3.37}$$

$$C = \frac{V_{c.p.}}{V'_{c.p.}} \tag{3.38}$$

For arrangement of road pavements and bases from FGC it is recommended to provide the necessary design tensile strength in bending of concrete ($f_{c.tf}$) using the following expression for calculating the W/C:

$$W / C = \frac{A f_{c.tf}}{f_{c.tf} + A \cdot 0.25 R_{c.tf}} \tag{3.39}$$

where $R_{c.tf}$ is cement tensile strength in bending, MPa; *A* is a coefficient that is equal to 0.5, 0.4 and 0.3 for coarse, medium and fine sand, respectively.

When calculating the FGC composition [31], the water demand method of determination (W) is taken based on the data, depending on the concrete mix workability, sand fineness, type and amount of plasticizers and air-entraining admixtures.

Consistently count:

Cement consumption (C), kg/m³:

$$C = W : (W/C) \tag{3.40}$$

Absolute volume of the cement paste ($V_{c.p.}$), l/m³:

$$V_{c.p.} = \frac{C}{\rho_C} + W \tag{3.41}$$

Sand content considering the tentative volume of entraining air ($V_{e.a}$):

$$S = [1000 - (V_{c.p.} + V_{e.a.})]\rho_s \tag{3.42}$$

At concrete mixture stiffness 10 … 20 sec and coarse or medium sand $V_{e.a} = $ 30 … 50 l/m³, for fine sand $V_{e.a} = 60 … 80$ l/m³. Then the cement consumption (C'), sand content (S') and water demand (W') for a test laboratory batch are obtained:

$$C' = \frac{m_b}{1 + n + W/C}$$

$$S' = \frac{m_b n}{1 + n + W/C}; W' = C' \cdot W/C \tag{3.43}$$

Where m_b is the laboratory batch mass; $n = S/C$.

After achieving the necessary workability by regulating the water demand and corresponding cement paste volume, the actual density of the compacted concrete mixture ($\rho_{c.m}$) is found and the entrained air volume is calculated:

$$V_{e.a} = [\frac{1 - \rho_{c.m}/(C') + S' + W')}{C'/\rho_c + S'/\rho_s + W}] \cdot 100\% \tag{3.44}$$

Factual components content per 1 m³ of sand concrete is determined using the following expressions:

$$V_{e.a} = 1000 - (C/\rho_C + S/\rho_S) \tag{3.45}$$

Selecting a methodology for design of FGC composition depends on the specific task conditions and features.

Granite crushing sifting is a special type of aggregate, which, apart from a certain size, is also characterized by a discontinuous grain composition, complex grain form and a surface relief, high voidness and presence of dispersed particles that act as a mineral filler. Such features of granite siftings cause a peculiar effect on the FGC characteristics, so the generally accepted coefficients in the strength equations require some refinement.

To determine the coefficients in Eq. 3.25, taking into account peculiarities of FGC on granite siftings, an approximation of compressive strength data and corresponding W/C at different values of technological factors was performed. Analysis of the experimental data allowed us to propose the average values of the coefficients A and *b* in Eq. 3.25 and Table 3.4.

Table 3.4 Coefficients A and *b* in Eq. (3.25) for calculating the fine-grained concrete strength.

Aggregate quality	Flowable concrete mixtures	Stiff concrete mixtures	Very stiff (semi-dry) concrete mixtures
High	A = 0.52, b = 0.65	A = 0.52, b = 0.55	A = 0.52, b = 0.2
Medium	A = 0.48, b = 0.65	A = 0.48, b = 0.55	A = 0.48, b = 0.2
Low	A = 0.44, b = 0.65	A = 0.44, b = 0.55	A = 0.44, b = 0.2

Based on the obtained models 3.32, a nomogram (Fig. 3.15) was constructed. This nomogram allows determining the cement consumption required for FGC on granite siftings, taking into account the C/W equation, concrete mixture workability, content of particles < 0.16 mm in the siftings and content of superplasticizer. Our proposed calculation method includes the following steps:

1. For given cement strength and required concrete strength value at 28 days, select the appropriate coefficients from Table 3.4 and use Eq. 3.25 to find the required C/W.
2. Calculate the W/C:

$$W/C = 1/(C/W) \tag{3.46}$$

3. Using the nomogram (Fig. 3.15) and considering the mixture fluidity, content of dust particles < 0.16 mm in siftings, find the cement consumption and superplasticizer content.
4. Knowing the cement consumption and the water—cement ratio, find the water demand:

$$W = C \cdot (W/C) \tag{3.47}$$

5. Knowing the cement content and density as well as the water demand, find the cement paste volume ($V_{c.p}$), aggregate volume (V_s) and mass (M_s) per 1 m³ of concrete mixture:

$$V_{c.p.} = \frac{C}{\rho_c} + W \tag{3.48}$$

$$V_s = 1000 - V_{c.p.} \tag{3.49}$$

$$M_s = V_s \cdot \rho_s \tag{3.50}$$

Numerical example

Let us calculate a FGC composition with a 28–day compressive strength of 40 MPa. Granite siftings are used as an aggregate with 24% of particles < 0.16 mm. The concrete mix cone slump (CS) is 10 cm. A plasticizing admixture is used, a naphthalene-formaldehyde type superplasticizer in a quantity of 0.6% by cement mass. In 28–days compressive strength of the cement is 50 MPa.

1. According to the given cement strength and required concrete strength value at 28 days, the C/W following Eq. 3.25 is

$$f_c^{28} = 0.44 \cdot R_c \left(C/W + 0.65 \right)$$

$$\frac{C}{W} = \left(\frac{f_c^{28}}{0.44 R_c} \right) - 0.65 = \left(\frac{40}{0.44 \cdot 50} \right) + 0.65 = 2.46$$

2. The corresponding value of W/C is

$$W/C = 1/2.46 = 0.41.$$

3. According to the nomogram (Fig. 3.15), the cement consumption is 460 kg/m³.
4. Following Eq. 3.47, for the known cement consumption and W/C, the water demand is

$$W = 460 \cdot 0.41 = 189 \ l/m^3.$$

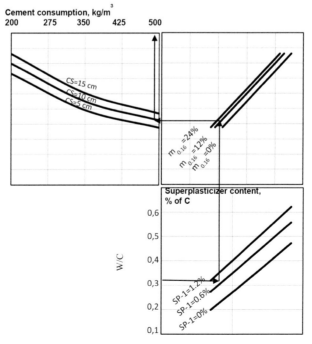

Fig. 3.15 Nomogram for obtaining cement consumption.

5. According to Eqs. (3.48–3.50), for known cement consumption and density as well as for known water demand, the cement paste volume, aggregate volume and mass are:

$$V_{c.p.} = \frac{C}{\rho_v} + W = \frac{460}{3.1} + 189 = 337 \ l;$$

$$V_{ag} = 1000 - V_{c.p.} = 1000 - 337 = 663 \ l;$$

$$M_{ag} = V_{ag} \cdot \rho_{ag} = 663 \cdot 2.7 = 1790 \ kg/m^3.$$

The calculated concrete composition is:

- cement – 460 kg/m³;
- siftings – 1790 kg/m³;
- water – 189 *l*/m³;
- superplasticizer SP – 2.76 kg/m³.

3.5 The effect of grain composition on FGC properties

The most common way to improve the properties of the aggregate on the basis of stone siftings is enrichment—the laundering of dust and clay particles. Due to the fact that laundering is a rather costly process, it is more practical to enrich the siftings by fractionation. This method is implemented by many stone processing enterprises. Fractionation allows for a separation of fine and coarse sand 0 … 0.53 mm and 0.63 … 2.5 mm from fine and coarse crushed stone fractions 2.5 … 5 mm and 5 … 10 mm. Siftings grain composition optimization by correcting the fractions content and simultaneous addition of superplasticizer should facilitate the selection of rational concrete compositions for different classes, including high-strength ones.

Experimental studies were carried out to investigate the effect of siftings' grain composition and using superplasticizers on the properties of concrete with high mixtures fluidity. Experiments were carried out using the "mix-technology-property" factor plan [51], allowing a simultaneous variation of the main aggregate fractions content and the concrete mixture composition (cement consumption and chemical admixture content). The experiments were carried out using granite siftings, Portland cement (C) with a standard strength of 59 MPa and naphthalene-formaldehyde superplasticizer.

The siftings were separated into three main fractions: 2.5 … 10 mm, 0.63 … 2.5 mm and 0 … 0.63 mm. The experiments planning conditions and results are given in Table 3.5.

According to the experimental plan, a FGC mixture with a cone slump of 16 … 21 cm, was prepared. The concrete mix water demand for the given cement consumption was selected in order to provide the given workability. Cubic specimens of 10 × 10 × 10 cm were prepared, which hardened under normal conditions and were tested at 28 days to find their compressive strength (f_c^{28}, MPa). Additionally, the aggregate

Table 3.5 Conditions for experiment planning corresponding to Eqs. 3.37 and 3.38.

Factors		Variation levels		
Natural	Coded	−1	0	+1
Content of crushed stone (2.5–10 mm), %	V_1	25	40	55
Content of coarse sand (0.63–2.5 mm), %	V_2	25	40	55
Content of fine sand (0–0.63 mm), %	V_3	20	35	50
Content of superplasticizer (%),	X_1	0	0.5	1
Content of cement (C kg/m³)	X_2	300	400	500

specific surface (S_{sp}, cm²/g) and voidness (V,%)—were controlled. After the factor plan realization experimental-statistical models 3.52, 3.53 were obtained, which allowed calculation of the corresponding iso-parametric diagrams, Figs. 3.16 ... 3.17.

$$W/C = 0.54V_1 + 0.6V_2 + 0.61V_3 - 0.17V_1V_2 - 0.02V_1V_3 -$$
$$0.11V_2V_3 - 0.2V_1X_1 - -0.05V_1X_2 - 0.2V_2X_1 -$$
$$0.07V_2X_2 - 0.2V_3X_1 - 0.08V_3X_2 + 0.05X_1^2 + 0.07X_2^2$$
(3.51)

$$f_c^{28} = 38.8V_1 + 25.3V_2 + 32V_3 + 20V_1V_2 - 7.6V_1V_3 -$$
$$27.5V_2V_3 + 12.1V_1X_1 + 1.7V_1X_2 + 6.6V_2X_1 - 1.5V_2X_2 +$$
$$8.9V_3X_1 + 3.2V_3X_2 - 0.1X_1X_2 - 4.6X_1^2 + 1.3X_2^2$$
(3.52)

Analysis of the models 3.51–3.52 shows that the grain composition factors mainly change the water demand in accordance with the change of the aggregate specific surface. An increase in fraction content of 2.5...10 mm reduces W/C by 0.03 ... 0.05 and increase in fraction of 0.63...2.5 increases the concrete mixture water demand. The influence of fraction 0...0.63 mm is ambiguous—with an increase in its content to

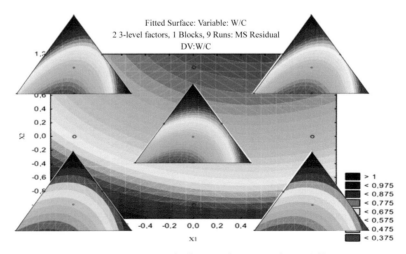

Fig. 3.16 Iso-parametric diagram of concrete mixture W/C.

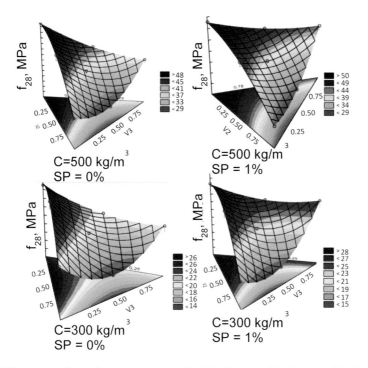

Fig. 3.17 Response surfaces of compression strength triple diagrams of grain composition influence at varied cement consumption and superplasticizer content.

40%, the water demand is reduced, and then there is a certain increase (Fig. 3.16). A small amount of fine fraction containing up to 40% of particles < 0.16 mm increases the mixture when leveling the negative effect on water demand due to superplasticizer. With further increase in granite sand content, the water demand increases.

The concrete compressive strength at 28 days varied from 18 to 50 MPa.

At constant values of other variable factors, the most influential factor is X_2 (cement consumption). There is also a positive effect of the superplasticizer (X_1), but the significant negative quadratic effect of this factor in the model 3.38, indicates the existence of an optimal value of X_1 within the variation range. The maximum increase in strength is observed when the admixture content is 0.4 ... 0.5% (Fig. 3.17). The range of maximum FGC strength practically corresponds to the minimum W/C, achieved at optimal aggregate grain composition (2.5 ... 10 mm –45 ... 55%; 0.63 ... 2.5 mm –25...40%; 0 ... 0.63 mm –20...35%). When the superplasticizer content increases, the negative effect of the disperse particles contained in the siftings decreases and, consequently, their rational amount increases.

The obtained mathematical models enable calculation of the most economical concrete compositions for different design classes.

3.6 FGC based on low-flowable mixtures

In order to study the effect of granite siftings for FGC made of low-flowable mixtures and to obtain the basic mathematical dependencies for concrete compositions design, a three-level type B_3 experimental plan was implemented. Experiment planning conditions are shown in the Table 3.6.

Table 3.6 Experiment planning conditions for Eqs. 3.39 … 3.45.

No.	Factors		Coded	Variation levels			Variation interval
	Natural			−1	0	1	
1	Cement consumption (C), kg/m³		X_1	200	350	500	150
2	Content of superplasticizer, % of C		X_2	0	0.5	1	0,5
3	Content of disperse particles in sifting, $m_{0.16}$, %		X_3	0	12	24	12

The concrete mixture fluidity at all the plan points was taken within the limits of 106–115 mm on the shake table, which corresponds to a standard cone slump of 1–2 cm. The investigated parameters of these experiments were taken of the concrete mixture W/C for achieving the specified fluidity and strength characteristics (compressive and bending strength at 3, 7, 28 days).

As a result of the experimental statistical processing the following mathematical models of the investigated parameters were obtained:

$$W / C = 0.52 - 0.18x_1 - 0.03x_2 + 0.05x_3 + 0.02x_1^2 + 0.01x_2^2 + 0.01x_3^2 - 0.02x_1x_2 - 0.03x_1x_3 - 0.01x_2x_3 \tag{3.53}$$

$$f_{c.tf}^3 = 6.79 + 2.07x_1 + 0.7x_2 + 0.04x_3 - 0.86x_1^2 - 0.1x_2^2 - 0.03x_3^2 + 0.1x_1x_2 - 0.49x_1x_3 + 0.06x_2x_3 \tag{3.54}$$

$$f_{c.tf}^7 = 8.71 + 2.75x_1 + 0.66x_2 + 0.03x_3 - 1.25x_1^2 - 0.48x_2^2 - 0.32x_3^2 + 0.33x_1x_2 - 0.76x_1x_3 - 0.16x_2x_3 \tag{3.55}$$

$$f_{c.tf}^{28} = 10.47 + 3.46x_1 + 0.83x_2 + 0.15x_3 - 0.74x_1^2 - 0.79x_2^2 - 0.69x_3^2 + +0.08x_1x_2 - 1.08x_1x_3 + 0.53x_2x_3 \tag{3.56}$$

$$f_c^3 = 30.30 + 10.68x_1 + 3.35x_2 - 0.18x_3 - 9.44x_1^2 - 0.05x_2^2 - 5.09x_3^2 - 0.21x_1x_2 - 2.75x_1x_3 + 0.32x_2x_3 \tag{3.57}$$

$$f_c^7 = 33.24 + 11.61x_1 + 2.66x_2 - 0.41x_3 -$$
$$7.72x_1^2 - 0.73x_2^2 - 2.67x_3^2 + 0.79x_1x_2 - \qquad (3.58)$$
$$5.47x_1x_3 + 0.59x_2x_3$$

$$f_c^{28} = 38.43 + 18.56x_1 + 3.10x_2 - 0.28x_3 -$$
$$7.48x_1^2 + 3.42x_2^2 - 0.88x_3^2 + 0.8x_1x_2 - \qquad (3.59)$$
$$5.5x_1x_3 + 1.05x_2x_3$$

Analysis of model 3.53 shows that in low flowable FGC mixtures on granite siftings (Fig. 3.18), as in high fluidity ones, the superplasticizer action yields a compensation of increasing water demand, caused by particles < 0.16 mm.

The FGC compressive strength within the variation range is 8 to 57 MPa. Mainly, the strength varied in accordance with the changes in those factors that cause the change in the concrete W/C, but the significant negative interaction effect of factors X_1 and X_3 draws attention—when these factors are at different variation levels there is a significant increase in concrete strength. This effect is caused by a decrease in the content of particles < 0.16 mm when increasing the cement consumption. This is due to the corresponding decrease in W/C and cement stone porosity (Fig. 3.19).

The positive effect of disperse granite filler becomes more noticeable in low fluidity mixtures with a small cement paste volume, which in some cases may not be sufficient to fill the voids of the sand fraction. This is especially noticeable in case of increased aggregate voidness, which is typical for granite siftings. With an increased content of superplasticizing admixture, an insufficient amount of cement paste causes significant concrete mix stratification.

The minimum W/C, achieved during the experimental plan implementation, is 0.32 … 0.35, and the compressive strength reaches 55 … 58 MPa, indicating that it is possible to produce high–strength FGC using siftings and superplasticizers.

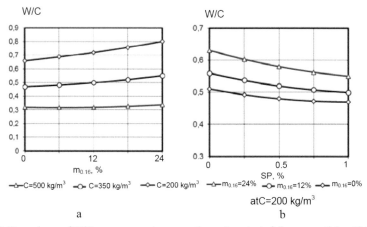

Fig. 3.18 Dependence of W/C on: a – cement consumption and content of disperse particles < 0.16 mm in siftings; b – content of disperse particles < 0.16 mm in siftings and superplasticizer content.

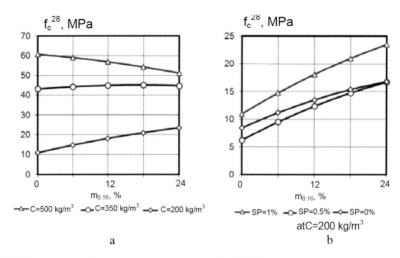

Fig. 3.19 Dependence of concrete compressive strength at 28 days on: a – cement consumption and content of disperse particles < 0.16 mm in siftings; b – content of disperse particles < 0.16 mm in siftings and superplasticizer content.

3.7 High strength FGC

Experiments were carried out to study the possibility of obtaining high-strength concrete on granite siftings, containing a significant amount of dusty grains, using effective chemical and mineral admixtures.

The experiments used Portland cement with standard compressive strength 50 MPa, polycarboxylate superplasticizers Melflux 2641F and Sika Viscocrete 225P (Table 3.7), as well as mineral admixtures with high-activity as metakaolin and microsilica. For correcting the grain composition in the siftings with $M_f = 3.23$ and content of particles < 0.16 mm of 17% sand fraction 2.5 ... 5 mm in amount of 20% was added. All concrete mixtures had the same aggregate to cement ratio (A/C = 3).The concrete mixture slump cone was within 9 ... 15 cm. Concrete compressive strength was determined at 3,7 and 28 days. The results are presented in Table 3.7.

The maximum compressive strength values of FGC on granite siftings, obtained in the experiments, were 85 ... 90 MPa. The most influential factors ensuring high strength

Table 3.7 Influence of admixtures on the strength of fine-grained concrete with granite siftings.

No.	Superplasticizer,%	Mineral admixture,%	W/C	CS, cm	Compressive strength (MPa), at		
					3 days	7 days	28 days
1	Melflux 2641F, 0.5%	-	0.32	12	63	71	78
2	Sika VC 225P, 0.5%	-	0.34	14	45	69	76
3	Melflux 2641F, 0.5%	metakaolin, 5%	0.37	12	43	48	53
4	Melflux 2641F, 0.5%	microsilica, 5%	0.35	13	40	56	62
5	Melflux 2641F, 1%	metakaolin, 5%	0.35	13	60	75	85
6	Melflux 2641F, 1%	microsilica, 5%	0.33	14	58	80	90

was the low concrete W/C (0.32 ... 0.37), which is achieved by using superplasticizers. Thus, using the Melflux admixture 2641f in amount of 0.5% by cement weight, the specified concrete mixture fluidity was achieved at W/C = 0.32, and in mixtures with Sika Viscocrete 225P (0.5%) at W/C = 0.34. The strength values at 28 days were, respectively, 78 and 76 MPa. Adding admixtures with high activity (metacolin and microsilica) at constant superplasticizer content (Melflux 2641F) increased W/C (up to 0.37 and 0.35, respectively). There was no increase in the concrete mixtures water demand caused by mineral additives with an increase in the superplasticizer content up to 1%. Increase of concrete strength caused by high activity mineral admixtures at low values of W/C, confirms the known data [58].

For quantitative estimation of the composition factors effect on the properties of high-strength FGC on granite siftings, algorithmic experiments were carried out according to the B_3 plan [59]. As variable factors have been chosen: content of polycarboxylate type superplasticizer Melflux 2641F (D, % (X_1)), content of particles < 0.16 mm in granite siftings ($m_{0.16}$,% (X_2)), content of metakaolin (MK,% (X_3)). Experiments planning conditions are given in Table 3.8. For the experiments siftings with an adjusted grain composition were used. The concrete composition, with exception of varying additives, was constant (C = 545 kg/m³, S = 1640 kg/m³). The following parameters were obtained: W/C of concrete mixture, required for reaching the cone slump (CS) 16 ... 21 cm and concrete compressive strength at 1 and 28 days (f_c^1 and f_c^{28}).

The regression equation for these parameters, depending on the variable factors values, is given below

$$W / C = 0.35 - 0.079x_1 + 0.02x_2 + 0.052x_3 + 0.026x_1^2 - 0.014x_2^2 + 0.011x_3^2 - 0.018x_1x_2 - 0.03x_1x_3 + 0.003x_2x_3 \tag{3.60}$$

$$f_c^1 = 2199 + 271x_1 + 0.34x_2 + 0.19x_3 - 0.828x_1^2 + 0.022x_2^2 - 2.37x_3^2 + 0.538x_1x_2 - 0.063x_1x_3 - 0.113x_2x_3 \tag{3.61}$$

$$f_c^{28} = 64 + 6.64x_1 + 2.7x_2 + 6.3x_3 - 1.1x_1^2 - 4.724x_2^2 - 8.326x_3^2 + 4.1x_1x_2 + 6.8x_1x_3 - 4.0x_2x_3 \tag{3.62}$$

As it follows from Eqs. 3.60 and 3.61, all variable factors cause significant changes in the concrete mixture water demand and, consequently, W/C and concrete strength. By the achieved effect, the factors can be sequenced as:

$$X_1(SP) > X_3(MK) > X_2(m_{0.16}).$$

The most significant interaction of factors in Eq. 3.60 is the "superplasticizer-metakaolin content" and "superplasticizer-particles < 0.16 mm content", which confirms the possibility of leveling the negative effect of the dispersion particles in siftings by superplasticizer at low W/C values (Fig. 3.20).

Table 3.8 Experiment planning conditions for equations (3.60–3.62).

No.	Factors		Variation level			Variation interval
	Natural	Coded	−1	0	1	
1	Content of superplasticizer Melflux 2641F (SP,%)	X_1	0	0.35	0.7	0.25
2	Content of particles < 0,16 mm in siftings ($m_{0.16}$,%)	X_2	0	6	12	6
3	Content of metakaolin (MK,%)	X_3	0	4	8	4

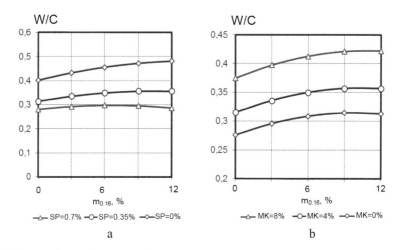

Fig. 3.20 Dependence of concrete W/C on: a – content of particles < 0.16 mm and addition of superplasticizer; b – content of particles < 0.16 mm and addition of Metakaolin.

Graphical dependencies Fig. 3.21, obtained by analysis of the corresponding experimental-statistical model, indicates the possibility of producing FGC on granite siftings that reaches maximum strength of 72 ... 75 MPa at 28 days. The maximum increase in strength is observed at optimal values of factors X_1 and X_2. At the same time, the effect of changing factor X_2. Is that content of particles < 0.16 mm within the variation limits is almost 2 times less than the superplasticizer content. Increasing the superplasticizer content (X_1) leads to an almost linear increase in strength, which, in general, corresponds to the nature change, achieved by the concrete mix W/C. For factors X_2 and X_3 which express the influence of dispersed mineral fillers, the existence of significant quadratic effects with a negative sign in the strength equations is characteristic, which shows the presence of the marginal region in their effective action. With the increase in the superplasticizer content, the effect of dispersed mineral components (granite dust and metakaolin) significantly increases. In compositions that contain no superplasticizer, increase of particles < 0.16 mm in the siftings to 4 ... 5 has a low effect on concrete strength. With further increase in the dusty particles content, the strength decreases by 23 ... 25%.

Adding the Melflux 2641F admixture contributes to the positive effect of granite filler, which causes the optimal combination of factors to increase the concrete strength

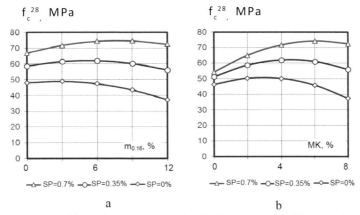

a   b

Fig. 3.21 Dependence of fine grain concrete strength at 28 days on superplasticizer quantity and content of particles < 0.16 mm in siftings (a), metakaolin (b).

by 12 ... 16%. The effective content of granite dust in the siftings increases with increase in superplasticizer content: if the content of the Melflux 2641F admixture is 0.35%, the positive effect is maintained up to 6 ... 7% of dust, with a maximum content of Melflux 2641F (0.7% by cement mass)—up to 8 ... 9%. Metakaolin (factor X_3), due to high dispersion and pozzolanic activity [59] is more effective than granite dust. The increase in the FGC strength at the expense of metakaolin is 35 ... 38%, which is consistent with the known data [59]. As for the dusty fraction of siftings, the effectiveness of metakaolin additive significantly increases with adding superplasticizer, although some positive effects (8 ... 10%) are observed in unplasticized mixtures.

The interaction effect of factors X_2 and X_3 that characterize the fillers content in Eqs. 1 ... 3, is negative, which indicates a decrease in their effectiveness at simultaneous increasing of the metakaolin content and that of granite particles < 0.16 mm. In this case, there is a significant increase in the mixture water demand, which cannot be neutralized by the steric effect of the polycarboxylate superplasticizer at its predetermined content.

For a maximum effect of metakaolin in producing high-strength FGC on granite siftings, the content of < 0.16 mm particles in the siftings should not exceed 4 ... 5%, which, in principle, is consistent with the requirements of the current standards [13]. The highest influence on achieving maximum early strength (at 1 day) has factor X_1 (the superplasticizer content), the influence of other factors is negligible and becomes evident just at maximum content of plasticizing admixture. Thus, subject to the use of effective polycarboxylate type superplasticizers and additional interaction of high-activity mineral admixtures, it is possible to obtain high-strength FGC using granite siftings as main aggregate. The obtained experimental-statistical models (3.46 ... 3.48) enable to propose a method for design of high-strength FGC compositions using granite siftings as aggregate.

To calculate the C/W, providing the specified concrete compressive strength at a certain age, a modified Eq. [31] that takes into account the influence of active mineral fillers can be used:

$$R_{cm} = AR_c \left((C/W)_{eq} - b \right)$$ (3.63)

where A is a coefficient, considering the aggregate quality, b is a coefficient that takes into account the concrete mixture type (Table 3.4), $(C/W)_{eq}$ is equivalent C/W of concrete, considering the mineral filler quantity and activity [31].

$(C/W)_{eq}$ is calculated as follows:

$$\left(C/W \right)_{eq} = \frac{C + K_{c.e.} \cdot F}{W}$$ (3.64)

Where C, W, F are cement consumption, water demand and mineral filler content, rrespectively, kg/m³, $K_{c.e.}$ is coefficient of mineral filler cementing efficiency.

As concrete mixture fillers are considered disperse components—active mineral admixtures and the content of dusty particles (< 0.16 mm) in siftings. Indicator $K_{c.e.}$ depends on the fillers characteristics (chemical and mineralogical composition, dispersion, hydraulic activity) and are usually determined experimentally. The obtained mathematical models of the strength for high-strength FGC on granite siftings with the use of metakaolin and effective polycarboxylate superplasticizer (Eqs. 3.47, 3.48) enable the calculation of the corresponding values of $K_{c.e.}$ (Table 3.9).

To determine the water demand in high-strength FGC a nomogram has been constructed (Fig. 3.22). The nomogram considers the influence of superplasticizer, active mineral admixture and particle content < 0.16 mm. The cement consumption is found as follows:

- if no mineral fillers are used

$$C = \left(C / W \right) \cdot W$$ (3.65)

- if mineral fillers are used

$$C = \left(C / W \right)_{eq} \cdot W - K_{c.e.} \cdot F$$ (3.66)

The aggregate content (granite siftings) is found from the equation of absolute volumes, taking into account the calculated of cement consumption and water demand values (Eqs. 3.34 ... 3.36).

Table 3.9 Calculated and experimental values of metakaolin and granite siftings cementing efficiency coefficients when using polycarboxylate superplasticizer.

Content of superplasticizer Melflux 2641F, %	Cementing efficiency coefficient of mineral fillers	
	Granite sifting dust	Metakaoline
0	−0.08	0.12
0.35	0.11	3.22
0.7	0.58	5.89

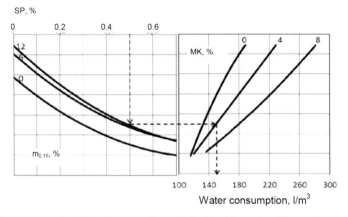

Fig. 3.22 Nomogram for finding the water demand of FGC with granite siftings (CS = 15–21 cm).

Numerical example. Calculate a composition of FGC with compressive strength corresponding to class C50/60, mixture cone slump 15 … 21 cm. Materials: Portland cement with standard compressive strength of 50 MPa, true density $(\rho_c) = 3.1$ g/cm³), granite siftings with dusty particles content of 10%, (true density $(\rho_3) = 2.7$ g/cm³)), superplasticizer Melflux 2641F.

1. Calculate the required average compressive strength of concrete at 28 days, which provides the concrete class C50\60 (with a standard coefficient of variation 13.5%). For this purpose, the well-known expression is used:

$$f_c^{28} = W \Big/ \left(1 - 1.64 \cdot \frac{C_v}{100}\right)$$

where C_v is variation coefficient, %.

$$f_c^{28} = 60 \Big/ \left(1 - 1.64 \cdot 0.135\right) = 77\ MPa$$

2. Considering the cement strength and the required concrete strength value at 28 days, according to Eq. (3.49) find the necessary $(C/W)_{eq}$. The coefficients in the equation are: $A = 0.44$, $b = 0.65$—as for low quality aggregate and corresponding mixture fluidity (Table 3.9).

$$f_{cm}^{28} = 0.44 R_c \left((C/W)_{eq} - 0.65\right)$$

$$(C/W)_{eq} = \left(\frac{f_c^{28}}{0.44 R_c}\right) + 0.65 = \left(\frac{77.1}{0.44 \cdot 50}\right) + 0.65 = 4.16$$

3. Using the nomogram (Fig. 3.22) find the water demand, taking into account the dusty particles content in siftings–W = 185 l/m^3.

4. Following Eq. 3.65, knowing the water demand and $(C/W)_{eq}$, find the cement consumption without taking into account the filler:

$$C = (C/W)_{np} \cdot W = 4.16 \cdot 185 = 768 \ kg/m^3$$

5. Following Eqs. (3.36, 3.37 and 3.41), find the cement paste volume and the aggregate weight:

$$V_{c.p.} = \frac{C}{\rho_c} + W = \frac{768}{3.1} + 185 = 433 \ l$$

$$V_a = 1000 - 433 = 567 \ l$$

$$A = 567 \cdot 2.7 = 1531 \ kg/m^3$$

6. Specify the high-strength FGC composition with the use the superplasticizer Melflux2641F and taking into account the disperse fraction of siftings as mineral filler:

   6.1 Find the approximate disperse fraction content in siftings (10% of the aggregate amount):

   $$F = 0.01 \cdot A = 0.01 \cdot 1531 = 153 \ kg$$

   6.2 Using the nomogram (Fig. 3.24), find the water demand, providing the required cone slump by selecting the effective superplasticizer quantity. At superplasticizer content of 0.5% by cement mass the required water demand is 135 l. From Table 3.9 find the cementing efficiency coefficient of the siftings' dusty fraction for the obtained superplasticizer content: $K_{c.e} = 0.35$.

   6.3 Taking the mineral filler content (F) to be 153 kg/m³, use Eq. 3.65 to calculate the specified cement consumption:

   $$C = (C/W)eq \cdot W - K_{c.e.} \cdot F = 4.16 \cdot 135 - 0.35 \cdot 153 = 508 \ kg/m^3$$

   6.4 Find the aggregate content:

   $$V_{c.p.} = \frac{C}{\rho_c} + W = \frac{508}{3.1} + 135 = 299 \ l$$

   $$V_a = 1000 - 299 = 701 \ l$$

   $$A = 701 \cdot 2.7 = 1893 \ kg/m^3$$

The calculated concrete composition per 1 m³ of concrete is:
cement 508 kg; aggregate (siftings) 1893 kg; water 135 l; superplasticizer 2.54 kg.

3.8 Complex plasticizing admixtures for producing high-strength concrete on granite siftings

Adding chemical admixtures and first of all superplasticizers, is one of the most effective ways for regulating the properties of concrete and reducing the cement consumption [13]. Recently, the most popular superplacticizers are high efficiency plasticizing admixtures based on polycarboxylate ethers. They differ from the known plasticizers by a higher water-reducing effect, allowing significant increase of density and improvement of strength and other concrete characteristics. A significant disadvantage of these plasticizers is their relatively high cost.

To reduce the cost and ensure the polyfunctional effect in concrete technology, complex admixtures are widely used. Such admixtures consist of several adjusting components, correcting and, in many cases, enhancing the effect of each of them. The purpose of this study was to develop a composition of complex admixtures based on the polycarboxylate ethers and other types of plasticizers, taking into account the content in cement of active mineral additive (blast granular slag) and water-cement ratio.

The research was carried out on FGC, in which as an aggregate was used a mixture of washed granite siftings fraction 0 ... 5 mm with quartz sand $M_f = 1.9$ in a ratio of 1 : 0.4. Portland cement with standard strength of 50 MPa and blast-furnace granulated slag, were used in the experiments. The slag was added into the cement during milling in a laboratory ball mill. As a plasticizing admixture were used polycarboxylate type superplasticizer Melflux 2651f (BASF, Germany), naphthalene-formaldehyde type superplasticizer and lignosulfonate type plasticizer LSTM. The cement consumption in the concrete mixture was constant (500 kg/m^3), W/C varied in the range of 0.35 ... 0.55. The tests were carried out on cubic specimens of 7×7 cm.

The work has included two stages: at the first stage the plasticization effect of individual admixtures and compositions based on them were studied and their effect, on strength at constant concrete mixtures water demand and different hardening durations were investigated. At the second stage the admixtures' water-reducing effect and its effectiveness to increase the concrete strength characteristics were studied.

The studies were performed using mathematical experiment planning. With this aim, algorithmic experiments were implemented according to the "composition-technology-properties" plan [50]. In this plan, simplex-planning of interconnected factors of components' mixture of and variation of independent technological factors are combined.

Factors that varied according to the pilot plan, in the first stage are:

- V_1 – content of plasticizer LSTM (0 ... 0.5%);
- V_2 – content of naphthalene-formaldehyde superplasticizer (SP), (0 ... 0.5%);
- V_3 – content polycarboxylate superplasticizer Melflux (0 ... 0.5%);
- X_1 – slag content (0 ... 30% by cement weight);
- X_2 – W/C (0.35 ... 0.55)

After statistical analysis of experimental data were obtained mathematical models for concrete cone slump (CS), the admixtures' plasticization effect (PE), concrete strength parameters in a form of the following polynomial regression equations:

$$
\begin{aligned}
y = {} & A_1V_1 + A_2V_2 + A_3V_3 + A_{12}V_1V_2 + A_{13}V_1V_3 + \\
& + A_{23}V_2V_3 + (Ab)_{11} V_1V_1 + (Ab)_{12} V_1V_2 + \\
& + (Ab)_{21} V_2V_1 + (Ab)_{22} V_2x_2 + (Ab)_{31} V_3x_1 + \\
& + (Ab)_{32} V_3x_2 + b_{12}x_1x_2 + b_{11}x_1^2 + b_{22}x_2^2
\end{aligned}
\tag{3.67}
$$

Models of the type 3.53 allow for carrying out calculations related with predicting the investigated output parameters that characterize concrete mixes and hardened concrete when adding the selected admixtures separately or in different compositions.

At the second stage of research, concrete mixtures were made at constant fluidity (CS = 15 ... 20 cm) using Portland cement without additives in order to obtain the required values of water-cement ratio and water-reducing effect (WRE) when changing the ratios between the selected admixtures in a wide range. Experiments were carried out using a simplex-grid Scheffe "mix-property" plan [50]. The mathematical models obtained by these experiments have the form:

$$
\begin{aligned}
y = {} & A_1V_1 + A_2V_2 + A_3V_3 + A_{12}V_1V_2 + A_{13}V_1V_3 + \\
& + A_{23}V_2V_3 + A_{123}V_1V_2V_3
\end{aligned}
\tag{3.68}
$$

The coefficients of the obtained mathematical models are given in Tables 3.10 and 3.11.

When varying the selected factors within the specified limits, the concrete mixtures cone slump changed from 4 to 26 cm. The highest effect on the concrete mixture fluidity have the content of admixtures and W/C.

Table 3.10 Coefficients of mathematical models of concrete workability (CS), plasticizing and water reducing effects (PE and WRE) of additives.

Coefficients	Output parameters			Coefficients	Output parameters	
	CS, cm	PE, %	WRE, %		CS, cm	PE, %
A_1	11.26	6	15.3	$(Ab)_{21}$	−4.64	−22
A_2	6.87	3	15.2	$(Ab)_{22}$	6.12	8
A_3	16.29	37	31.3	$(Ab)_{31}$	−0.53	−6
A_{12}	−11.99	−46	17.4	$(Ab)_{32}$	5.59	5
A_{13}	10.96	88	37.4	b_{12}	−1.38	−9
A_{23}	3.47	0	20	b_{11}	−0.48	3
A_{123}	-	-	41.8	b_{22}	3.23	23
$(Ab)_{11}$	−0.79	−5	-			
$(Ab)_{12}$	7.25	9	-			

The effectiveness of the factors complex influence on the mixture fluidity was estimated by the plasticization effect magnitude, which was obtained as follows:

$$PE = \frac{CS - CS_0}{CS_0} \cdot 100\% \qquad (3.69)$$

Where CS is the cone slump at adding plasticizing admixture; CS_0 is the cone slump for the control concrete composition.

The plasticization effect (PE) of the investigated admixtures with different composition and at a maximum their content of 0.5% by cement weight was in the range from 22% to 82%. As follows from the PE mathematical models (Table 3.11), the highest plasticization effect has been obtained due to adding by the polycarboxylate type Melflux superplasticizer (with no slag in cement; PE varies from 50% to 82% for concrete mixtures with W/C = 0.35 and W/C = 0.55; if the a slag content in cement is 30% PE varies from 38% to 72% for W/C = 0.35 and W/C = 0.55 respectively. The LSTM plasticizer and the superplasticizer SP showed a slightly lower plasticization effect value (Fig. 3.23). The plasticization effect of all investigated admixtures increases with increasing W/C.

Adding milled blast furnace slag causes a certain decrease in the plasticization effect of the studied admixtures. This is more pronounced at W/C values from 15 to 30% (Fig. 3.23b), with low W/C the effect of slag on the plasticization effect of the admixture is insignificant (up to 15%).

Table 3.11 Mathematical models coefficients for concrete strength parameters.

Coefficients	Output parameters				Coefficients	Output parameters			
	Compressive strength (f_c), MPa at					Compressive strength (f_c), MPa at			
	12 h	1 day	7 days	28 days		12 h	1 day	7 days	28 days
A_1	1.79	3.6 / 9.4*	27.9 / 30.7	34 / 39.6	$(Ab)_{21}$	−0.41	−0.12	−1.88	−3,41
A_2	6.74	13.5 / 18.9	23.7 / 43.1	35.4 / 46.7	$(Ab)_{22}$	−0.90	−4.36	−1.44	−2.80
A_3	6.85	13.7 / 18.6	40.5 / 54.3	46.8 / 61.9	$(Ab)_{31}$	−0.10	−1.18	1.72	−5.22
A_{12}	3.28	6.6 / −4.8	−23.4 / 15.1	49.6 / −5.8	$(Ab)_{32}$	−0.90	−3.33	−2.55	−1.00
A_{13}	2.47	4.9 / 16.3	−9.4 / −42.2	33.1 / −6.1	b_{12}	−0.30	−0.04	−0.05	10.64
A_{23}	−2.42	−4.9 / 2.9	0.01 / 41.9	22.7 / 43.9	b_{11}	0.18	0.67	1.51	0.43
A_{123}	-	−/−4	−/106	−/176	b_{22}	0.14	0.30	0.71	−3.41
$(Ab)_{11}$	−0.55	−1.09	−5.73	2.46					
$(Ab)_{12}$	0.00	0.00	−1.08	−4.83					

*above the line – coefficients of mathematical models for concrete strength without considering the water reducing effect, under the line – considering the water reducing effect.

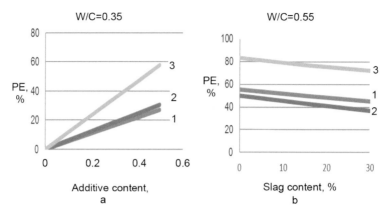

Fig. 3.23 Influence of ordinary additives on plasticizing effect (PE), % 1–LSTM; 2 – SP; 3–Melflux.

When evaluating the joint effect of admixtures of different types, the maximum plasticization effect was observed when LSTM plasticizer was combined with a Melflux superplasticizer in the same ratios (Fig. 3.24a).

In this case, the PE was within the range of 52 ... 82%, which is close to the effect obtained with adding only the Melflux admixture. The combined Melflux and naphtahlene-formaldehyde superplasticizer admixture SP leads to a decrease in PE to 35 ... 68%. The lowest plasticization effect value was obtained by combining the LSTM plasticizer and the superplasticizer (SP).

Combination of the three investigated admixtures in the same proportions enables to obtain PE of 60 ... 63% at W/C = 0.55 and PE of 29 ... 31% at W/C = 0.55. Analysis of the joint effect of admixtures using the triple diagram allowed to find the correlation region, providing the maximum plasticization effect at W/C = 0.35 using 68 ... 73% of Melflux, 8 ... 12% of SP and 18 ... 23% of LSTM yields PE = 37 ... 45%; At W/C = 0.55, 48 ... 53% of Melflux-, 8 ... 15% of SP and 35 ... 42% of LSTM PE = 55 ... 65%).

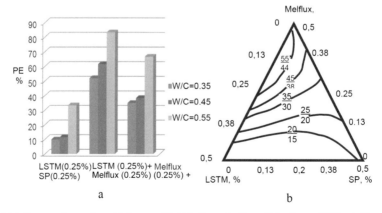

Fig. 3.24 Influence of complex additives on PE, % * Above the line – PE values at W/C = 0.55, under the line – PE values at W/C = 0.35.

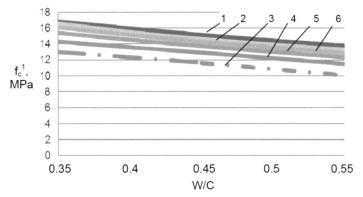

Fig. 3.25 Influence of additives at their total content 0.5% of cement on concrete compressive strength (f_c^1) at 1day (without considering the water reduction effect): 1 – SP , 2 – Melflux, 3 – concrete without admixtures, 4 – LSTM+SP 1, 5 – LSTM+Melflux, 6 – SP+Melflux (in complexad mixtures components have equal ratio by mass).

The results of research show that concrete with SP and Melflux admixtures, substantially increase the concrete mixtures fluidity, and also have a positive effect on strength even under unchanged W/C values (Fig. 3.25). Concrete strength values at 12 h and 1 day without taking into account the water-reducing effect significantly decrease when LSTM admixture is added. The negative effect of LSTM admixture is practically not evident at 7 days; at 28 days there is even some positive effect. The negative effect of LSTM admixture in early age is offset by combination of this plasticizer with superplasticizer SP as part of complex admixtures. Less effective are LSTM and Melflux compositions.

At 28 days, at constant W/C values, the influence of complex admixtures, including LSTM and SP, LSTM and Melflux, SP and Melflux, is practically equivalent. Even without taking into account the water-reducing effect, all these admixtures increase the strength by 20 ... 25%. The highest strength increasing effect is observed at low W/C values. Combination of three admixtures—Melflux, LSTM and SP leads to a certain decrease in their positive effect at early age strength of concrete at constant W/C values (Fig. 3.26).

The positive influence of complex admixtures on concrete strength significantly increases when taking into account their water-reducing effect. The latter can be calculated as follows:

$$WRE = \frac{W - W_0}{W_0} \cdot 100\% \qquad (3.70)$$

where W is water demand at adding plasticizing admixture without reducing the initial concrete mixture mobility (CS = 16 ... 20 cm); W_0 is the water demand for concrete mixture without plasticizer.

The water-reducing effect (WRE) of the investigated admixtures varied within 13 ... 28% admixture. The best WRE was obtained by the adding of the admixture

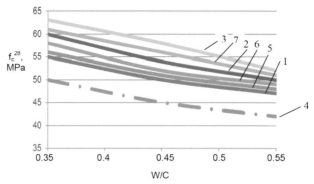

Fig. 3.26 Influence of admixtures at their total content of 0.5% by cement mass on compressive strength (f_c^{28}) at 28 days (considering water reduction effect). 1 – LSTM 2 – 3 – Melflux, 4 – concrete without admixtures, 5 –LSTM+SP, 6-LSTM+Melflux, 7 – SP+Melflux, (in complex admixtures components have equal ratio by mass).

based on polycarboxylate ether (25–30%), lower WRE was achieved by adding of admixtures based on lignosulfonate and naphthalene formaldehyde (10 ... 15%). When evaluating the joint effect of different admixtures types, the best VRE is observed when combining the Melflux superplasticizer and the plasticizer LSTM in the same ratios. In this case the WRE is within 24 ... 28%, which is close to the effect obtained with adding Melflux admixture individually. Melflux and SP as well as the three components admixture, including Melflux, SP and LSTM, showed a slightly lower WRE. Analyzing the joint effect of admixtures on a triple diagram, it is possible to obtain the limits of admixtures that provide the maximum WRE: Melflux –50 ... 55%, SP-10 ... 15%, LSTM-30 ... 35% (Fig. 3.27).

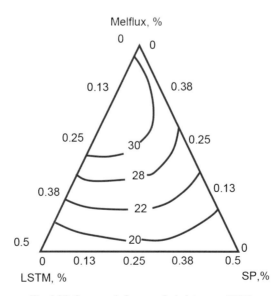

Fig. 3.27 Common influence of admixtures onWRE.

Direct relationship between the increase in strength and WRE is observed when changing the content of a specific admixture in concrete. The general relationship between the strength gain at one day (Δf_{c1}), and at 28 days (Δf_{c28}) and WRE is not unambiguous for concrete with different types of admixtures (Fig. 3.29). The strength gain of double complex SP and Melflux admixtures, as well as triple ones (LSTM, SP, Melflux) is closer to that obtained with the adding an individual Melflux admixture. The best admixtures that provide the maximum strength gain due to their water-reducing effect are given in Table 3.12.

Finding the optimal composition of complex admixture depends on the specific optimization conditions. Such conditions can be: for example, ensuring maximum concrete strength at 1 day with constant concrete mix fluidity or ensuring maximum concrete mix fluidity without reducing the early strength, etc.

Figure 3.29 shows an example of a graphically-analytical selection of the optimal complex admixture composition that provides concrete compressive strength values of at least 12 MPa at 1 day, 60 MPa at 28 days and at least 20% for the plasticization effect. The region of complex admixtures compositions, providing the necessary levels of the given parameters, was calculated using mathematical models. As it follows from Fig. 3.29, this region is: Melflux –25 ... 35%, SP-40 ... 50%, LSTM-25 ... 35.

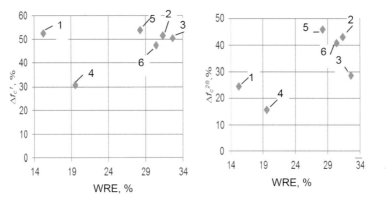

Fig. 3.28 Dependences of strength change (Δf_{cm}) at one (a) and 28 (b) days on WRE, % 1 SP; 2 – Melflux; 3 – LSTM + Melflux; 4 – LSTM + SP; 5 – SP + Melflux; 6 – LSTM + SP + Melflux. The total content of admixtures 0.5% by cement mass (in complex admixtures components have equal ratio by mass).

Table 3.12 Concrete mix strength increase value considering WRE.*

Admixture	Ratio, by mass	WRE, %	Δf_c^1, %	Δf_c^{28}, %
SP	-	15.22	52.49	24.55
Melflux	-	31.30	51.72	43.18
LSTM+SP	1:1	19.57	30.78	15.60
LSTM+Melflux	1:1	32.61	50.44	28.57
SP+Melflux	1:1	28.26	53.87	46.08
LSTM +SP+Melflux	1:1:1	30.43	47.50	40.86

* total admixture content 0.5% by cement weight.

The obtained mathematical models of concrete strength, considering the effect of plasticizers, enable to propose calculation dependences for selection of C/W for design the compositions of concrete mix with admixtures. Analysis shows (Fig. 3.30) that when adding admixtures there is still linear dependence of concrete strength on C/W.

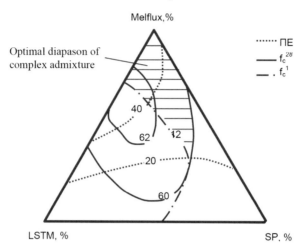

Fig. 3.29 Example of selecting a region of optimal complex plasticizing admixture composition.

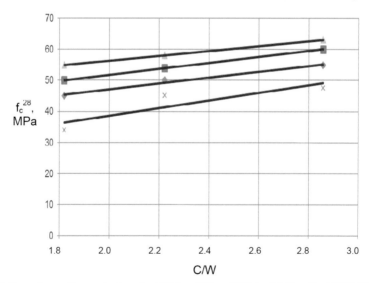

Fig. 3.30 Dependence of concrete strength at 28 days (f_c^{28}) on C/W. 1 – Melflux, 2 – LSTM (0.166%) + SP (0.166%) + Melflux (0.166 %), 3 – SP (0.5%), 4 – LSTM (0.5%).

When approximating experimental data, the calculated concrete strength expression can be represented as a general dependence, which is consistent with the known recommendations [31]:

$$f_c = kAR_{\text{ц}}\left(\frac{C}{W} + b\right) \tag{3.71}$$

where k is coefficient taking in to account the admixture type, A, B are coefficients, considering the studied parameters' quality, concrete age, hardening conditions, f_c is strength of cement, MPa, C/W is the cement – water ratio.

Table 3.13 presents the calculated values of coefficients k and b (A = 0.44).

Table 3.13 Values of coefficients in the strength Eq. 3.71.

Plasticizing admixture	Equation coefficients	
	k	b
Melflux (0.5%)	0.43	2.98
SP (0.5%)	0.44	3.4
LSTM (0.5%)	0.35	5.22
LSTM (0.166%) + SP (0.166%) + Melflux (0.166%)	0.55	1.16

The performed research has proved the possibility of rational combination of polycarboxylate type superplasticizers with plasticizers of other types and the creation of effective complex admixtures, characterized by high plasticizing and water-reducing effects. Analysis of polynomial models obtained using the "composition-technology-properties" plans allows performing the necessary calculations to optimize the complex admixtures composition and find the basic parameters of concrete mixtures composition with these admixtures.

CHAPTER 4

Vibro-Pressed Fine Grained Concrete Based on Stone Siftings

4.1 Technological background

To obtain quality products from stiff mixtures, it is necessary to provide maximum compacting, and compacting degree is one of the main criteria for determining the mechanical properties of concrete. The final aim of obtaining a most dense concrete from stiff mixtures is complete removal of involved air.

If the mixture is fully compacted and high quality aggregate is used, the strength of concrete from stiff mixtures at any hardening conditions, practically linearly, depends on the water-cement ratio [60]. At constant water-cement ratio the concrete strength increases with concrete mix stiffness. This tendency is especially noticeable in concrete with high water-cement ratio and at early hardening stages [61].

Increase in concrete mixtures' stiffness is associated with an increase in aggregate volume and decrease in that of cement paste. In this case concrete strength is determined more by the constructive role of the aggregate. The aggregate carcass takes the loading therefore the cement stone strength has lower influence. That is why the effect of increasing the compressive strength with increasing the concrete mixture stiffness is more noticeable with relatively low cement stone strength. With its high strength (when it is close to that of the aggregate), the effect of the stone carcass is not so significant, and with increasing the stiffness, concrete strength increases a little [60].

The strength of concrete from stiff mixes at the same amount of cement stone and strength, it decreases with an increase in the content of the cement-sand component of the concrete mixture, that is, with an increase in the grains' extension coefficient, α [60].

According to the physical-chemical theory [62], the density and strength of the hydrated shells on the grains of particles of aggregates is reduced with the distance from the aggregate surface. In stiff concrete, when the aggregate particles approach the minimum distance [28, 60], the density in the contact area with the aggregate is higher than that of the cement stone in general.

Due to the relatively low water-cement ratio values, concrete from stiff mixtures hardens faster, especially at the early stages. The low water demand of such concrete is due to the fact that it has a fine porous structure that holds water, especially when heated. This creates additional favorable conditions for the heat treatment process, allows accelerating the rise of temperature, increases the temperature of isothermal aging and reduces the medium humidity distance [60, 28].

Stiff concrete mixtures require intensive compacting methods. Just moderately stiff mixtures can be densely compacted by ordinary vibration (under the condition of optimally chosen amplitudes and frequencies) [60]. With an increase in the vibration amplitude to 0.7 ... 0.9 mm, it is possible to achieve compacting of especially stiff mixtures, but the duration of vibration is significantly increased. Mixture with high stiffness cannot be completely compacted even at high vibration amplitude.

Vibro-pressing is one of the most effective and relatively technically simple compacting methods of concrete mixtures with high stiffness. This method of concrete products forming has two main types:

- compacting by simultaneous vibration and pressure (vibration with loading);
- combination of vibration compacting with subsequent static pressing.

The expediency of force compaction of cement systems containing disperse filler is confirmed by considering the processes occurring at the level of physical interaction between the filler particles, both with each other, and with the particles of hydrated cement. The filler particles are held in the spatial cement net by coagulation forces [18]. Intensive compacting methods lead to a decrease in the distance between particles that, as it follows from Eq. (2.1), contributes to the increase of the coupling force between them and yields the Coulomb forces, providing the maximum contact strength, which is unattainable in systems compacted without power methods.

As known, possibility of combining adding fillers and vibro-pressing has been experimentally confirmed. Positive influence of milled sand on the properties of concrete compacted by vibration and vibro-pressing was noted by Ahverdov [28]. It is shown that using it in fine-grained concrete such fillers as loam and crushed ground with a specific surface of 200 ... 300 m^2/kg yields cement saving up to 30 ... 40 kg / m^3 in vibro-pressed concrete production at P = 0.015 MPa.

Effectiveness of adding a filler into stiff concrete mixtures compacted by vibro-pressing (P = 0.01 MPa) was also reported for a fine-grained perlite-concrete mixture with fly ash to ensure better products molding. As a result of adding fly ash, the cement consumption dropped by 30 ... 50%, while the effect of vibro-pressing, compared usual vibration compacting, resulted in a density increase of 3–6% and compressive strength growth.

Along with traditional fillers from industrial waste (fly ash, milled slag, microsilica), use of rocks crushing and processing waste is possible. Improvement of physical and mechanical properties of cement composites based on finely dispersed stone treatment waste (50% moisture sludge, represented predominantly by calcium carbonate) was demonstrated. It was noted that disperse fillers are a substrate for the hydrated compositions at structure formation of cement stone, possibly also epitaxial merging of calcite and portlandite crystals and chemical interaction of calcite with belite.

4.2 Influence of granite filler grain composition on formation of vibro pressed fine-grained concrete structure

The role of a dispersed mineral filler in hardening cement systems is largely determined by its grain composition. If the filler grain composition goes beyond the optimum, there is a tensile force on the whole cement stone volume or on its individual sections. In order to avoid this force it is necessary to include in the filler composition a certain number of grains of a larger fraction that could form contacts in the hardening system, caused by electric attraction or mechanical clamping forces [63]. Prerequisites for the occurrence of such contacts are additionally created during power compacting of ultra-stiff concrete mixtures, when the distance between the interacting particles decreases, facilitating their mutual attraction.

In the present research a simplex-grid plan [50] was used. The filler was represented by a mixture of three granite siftings fractions: $V_1 - 0.315 ... 0.16$ mm, $V_2 - 0.16 ... 0.08$ mm, $V_3 - < 0.08$ mm. Cylindrical specimens $d = h = 50$ mm were made by the vibration method with loading $P = 0.06$ MPa and compacting time $\tau = 15$ seconds. The concrete composition at point number 1 (Table 4.1) with the least concrete mixture water demand was as follows: C (cement) = 300 kg/m³, A (aggregate) = 1750 kg/m³, W (water) = 156 l/m³. After finding the required water demand, the concretes compositions was adjusted to maintain a constant W/C (W/C = 0.52) when changing the filler fraction ratio.

Due to a significant change in the concrete mixture water demand, depending on the filler grain composition and the associated significant fluctuation in cement consumption at different factor plan points, to enable a more objective analysis of the results, along with the basic output parameters (compressive strength at 28 days f_c^{28}, MPa, the open porosity P_v, %, the average pore size λ and the pore size uniformity α_p), the cement efficiency coefficient was determined: $k_e = \dfrac{f_c}{C}$, MPa·m³/kg. It characterizes the concrete strength from 1 kg of cement per 1 m³ of concrete. Experiments were duplicated at all points at a constant of cement consumption and aggregate content. The results of this investigation are given in Table 4.1.

As a result of the experimental data, statistical processing adequate mathematical models of the studied parameters were obtained and their triple diagrams were constructed (Figs. 4.1, 4.2):

at W/C = const:

$$f_c^{28} = 28.9V_1 + 30.7V_2 + 38.8V_3 - 2.3V_1V_2 + 13.5V_1V_3 - 13.4V_2V_3 + 0.9V_1V_2V_3 \tag{4.1}$$

Table 4.1 Influence of the filler dispersion on vibro-pressed fine grained concrete (VFC) properties.

No.	Filler Grain composition,%			Concrete composition			Studied (output) parameters' values				
	V_1	V_2	V_3	C, kg/m³	W, l/m³	W/C	f_c^{28}, MPa	$P_v\%$	a_n	λ	k_e, MPa·m³ kg
1	100	0	0	300	156	0.52	28.9	16.54	0.55	2.42	0.096
				300	156	0.52	28.9	16.54	0.55	2.42	—
2	0	100	0	378	197	0.52	30.7	15.11	0.52	2.25	0.081
				300	184	0.61	36.9	14.7	0.5	2.23	—
3	0	0	100	394	205	0.52	38.8	9.59	0.71	1.74	0.098
				300	189	0.63	39.2	10.1	0.67	1.81	—
4	50	50	0	304	158	0.52	29.2	15.56	0.53	2.35	0.096
				300	157	0.52	32.9	15.32	0.51	2.33	—
5	50	0	50	339	176	0.52	37.2	14.67	0.68	2.01	0.110
				300	170	0.57	36.4	13.83	0.65	2.21	—
6	0	50	50	325	169	0.52	31.4	13.35	0.6	1.97	0.097
				300	165	0.55	37.1	12.1	0.56	2.03	—
7	33,3	33,3	33,3	326	170	0.52	35.6	14.82	0.64	2.12	0.107
				300	165	0.55	36.0	13.68	0.61	2.19	—

Notes: 1. The filler content is 25% of the aggregate weight 2. Above the line are given experimental values, obtained at W/C = const, under the line – at C = const.

$$P_v = 16.5V_1 + 15.1V_2 + 9.6V_3 - 1.1V_1V_2 + 6.4V_1V_3 + 4V_2V_3 + 0.9V_1V_2V_3 \tag{4.2}$$

$$a_p = 0.55V_1 + 0.52V_2 + 0.71V_3 - 0.02V_1V_2 + 0.2V_1V_3 - 0.06V_2V_3 + 0.9V_1V_2V_3 \tag{4.3}$$

$$\lambda = 2.42V_1 + 2.25V_2 + 1.74V_3 + 0.06V_1V_2 - 0.28V_1V_3 - 0.1V_2V_3 + 0.51V_1V_2V_3 \tag{4.4}$$

$$k_e = 0.096V_1 + 0.081V_2 + 0.098V_3 - 0.028V_1V_2 + 0.049V_1V_3 - 0.027V_2V_3 + 0.11V_1V_2V_3 \tag{4.5}$$

at C = const:

$$f_c^{*28} = 28.9V_1 + 36.9V_2 + 39.2V_3 + 11.2V_1V_3 - 3.7V_2\ V_3 + 2.8V_1V_2V_3 \tag{4.6}$$

$$P_v^* = 16.5V_1 + 14.7V_2 + 10.1V_3 - 1.2V_1V_2 + 2V_1V_3 - 1.2V_2V_3 - 1.6V_1V_2V_3 \tag{4.7}$$

$$a_p^* = 0.55V_1 + 0.5V_2 + 0.671V_3 - 0.06V_1V_2 +$$
$$0.16V_1V_3 - 0.1V_2V_3 + 0.99V_1V_2V_3 \tag{4.8}$$

$$\lambda^* = 242V_1 + 2.23V_2 + 1.81V_3 + 0.02V_1V_2$$
$$0.37V_1V_3 + 0.04V_2V_3 - 0.33V_1V_2V_3 \tag{4.9}$$

Analyzing the obtained regression equations, it should be noted that in the case when the cement consumption was maintained at a constant (Eq. 4.6), increasing the filler dispersion to the maximum variable leads to an increase of f_c^{28} by 33 ... 35%, although the W/C increases by 25 ... 27 % (Fig. 4.1). The highest change in strength is observed when grain sizes change from 0.315 ... 0.16 to < 0.08 mm. Similarly, but more smoothly the strength increases during the transition from fraction 0.315...0.16 to 0.16 ... 0.08 mm. Increasing the filler dispersion from 0.16 mm also causes an increase of f_c^{28}, but only by 7 ... 8%. Thus, despite the significant increase in the concrete mixture water demand, increasing the granite filler dispersion to particles < 0.08 mm causes a clear increase in compressive strength, decrease in the open porosity, as well as improvement of porosity parameters (reducing the average size and increasing the pores uniformity).

Although for W/C = const the effect of individual fractions on strength is similar, the general nature of the filler particle size effect is slightly different. An increase in the content of fraction < 0.08 mm in a mixture with a fraction of 0.315 ... 0.16 mm increases the strength, but with the content of this fraction more than 50 ... 60% of the mixture mass, the strength increases insignificantly. Combination of fractions 0.315 ... 0.16 mm and 0.16 ... 0.08 mm yields an evident effect of strength increasing after the content of the last fraction grows to 60 ... 70%. The strength dependence on mutual influence of fractions < 0.08 and 0.16 ... 0.08 mm has a character of a progressive increase in the direction of dispersion growth.

If for C = const the most effective is the increase of the granite filler dispersion, then for W/C = const, as shown by the mathematical model k_e (4.5), such a method is not the most optimal, because the resulting increase in cement consumption is not sufficiently compensated by increased strength. Optimal combination of maximum f_c^{28} and expedient cement consumption is obtained by using a mixture of fractions: < 0.08 mm - 35 ... 50%, 0.16 ... 0.08–10 ... 40%, 0.315 ... 0.16–40 ... 60%. At W/C = const, the porosity parameters change in the same way as for C = const, but the region of their optimal values (Fig. 4.2) does not coincide with the optimum k_e and f_c (Fig. 4.1), due to the presence of a large number of big grains, the structure becomes less homogeneous.

The step that is of the highest importance in the process of the cement stone formation with filler has an initial stage during which the coagulation structure formation takes place. Its strength is directly related to that of the final (crystallization) structure [18]. The filler influence on this process can be estimated quantitatively and qualitatively from the kinetics of cement stone and concrete time-varying structural and mechanical parameters.

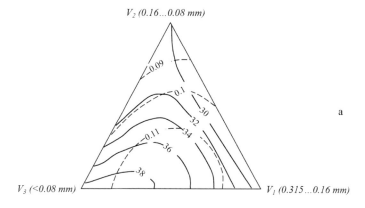

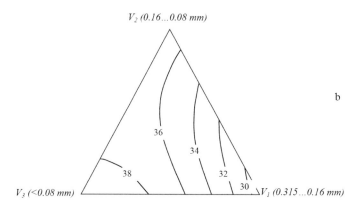

Fig. 4.1 Strength (f_c^{28}, MPa (-------) and cement efficiency (k_e, MPa·m³/kg (- - - -): (a) at W/C = const; (b) at C = const.

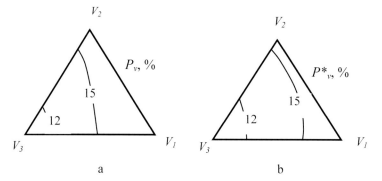

Fig. 4.2 Influence of filler granulometric composition on vibro-pressed fine grained concrete porosity: (a) W/C = const; (b) C = const.

Kinetics of electrical conductivity and ultrasonic waves' passing velocity was obtained using VFC specimens. Kinetics of water chemical binding in a cement stone, depending on the dusty fraction content in granite siftings was also studied. Compositions for preparing concrete samples for measuring the electrical conductivity and the ultrasonic waves velocity are given in Table 4.2.

As a result of measuring the change in electrical conductivity of the filled cement stone in time, a number of curves that have a rather characteristic shape were obtained (Fig. 4.3). During the first 10 ... 20 min. after the compacting of the concrete mixture there is a slight decrease in electrical conductivity (α), which is then quite abruptly terminated and remains almost constant for 1 ... 2 hours (depending on the concrete composition of the specimen). After that the electrical conductivity begins to increase and in a few hours it reaches its maximum conductivity. Furthermore there is a monotonic decline α and only on some curves relating to concrete, in which there is no filler of small fractions, there is another, but less significant maximum conductivity.

As known [64], the dependence curves of the ultrasonic waves velocity passing through concrete C_l from time contain areas characterizing the structure formation stages: in the initial stage (induction period) C_l slightly increases (horizontal section is noted). Further, due to the formation of low-strength, but significantly influencing on C_l crystallization frame of cement aluminate components $\dfrac{dC_l}{d\tau}$ becomes maximal. With the beginning of hydrosilicates crystallization, which are the main strength carriers, C_l growth rate decreases. As it follows from our experiments, for VFC the initial horizontal section is not characteristic—the change in the ultrasonic waves passage velocity occurs smoothly with the transition to the highest velocity $\dfrac{dC_l}{d\tau}$ (Fig. 4.4).

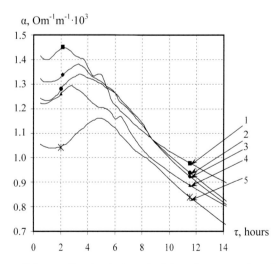

Fig. 4.3 Dependence of the specific electrical conductivity of concrete α) on hardening duration: 1 – dust fraction content (< 0.16 mm) – $m_{0.16}$ = 0%, C = 300 kg/m³, A = 1750 kg/m³, W = 118 l/m³; 2 – No. 1 from Table 4.1; 3 – dust fraction content – $m_{0.16}$ =18%; C=300 kg/m³, A = 1750 kg/m³, W = 160 l/m³, 4 – No. 2 from Table 4.1; 5 – No. 3 from Table 4.1.

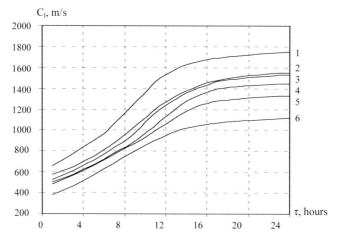

Fig. 4.4 Dependence of the longitudinal ultrasonic waves passage velocity (C_l) on hardening duration: 1 – W/C = 0.4, $m_{0.16}$ = 18%; 2 – W/C = 0.67, $m_{0.16}$ = 18%; 3 – W/C = 0,4, $m_{0.16}$ = 10,5%; 4 – W/C = 0.4, $m_{0.16}$ = 0%; 5 – W/C = 0.67, $m_{0.16}$ = 10,5%; 6 – W/C = 0.67, $m_{0.16}$ = 0%.

It is obvious that after the vibro-pressing of super stiff concrete mixtures a dense initial concrete structure is formed, which is characterized by considerable strength compared with concrete from plastic mixtures, and therefore the coagulation contacts formation in "compressed" conditions leads to increase in the system elasticity, which is reflected in the rate of C_l growth. In concrete with filler the hydrosilicate compositions crystallization and the associated increase in strength go faster.

Investigations of structure formation features for VFC on granite siftings by the ultrasonic waves velocity of (C_l, m/s) confirms the positive effect of granite filler on the structure formation at early age. It is noticed that the general rate of the longitudinal ultrasonic waves velocity growth $\dfrac{dC_l}{d\tau}$ increases with adding 18 ... 20% of granite powder at W/C of 0.4 – by 8 ... 9%, while at W/C = 0,67 – by 38 ... 40%. It also confirms the greater influence of granite filler in low–cement concrete.

To study the granite filler influence on vibro-pressed cement stone hydration degree, cylindrical specimens $d = h = 2,5$ cm were made which were compacted for 15 seconds under a load of 0.06 MPa. The specimens have hardened in air-humid conditions, after 3, 7, 14, 28 days. The amount of chemically bounded water and the Portland cement hydration degree were obtained. The results of experiments are shown in Fig. 4.5.

As it can be seen from Fig. 4.5, vibro-pressed cement stone without a dusting filler, hydrates to a lesser extent than that with a filler. On the third day the hydration degree (α) for cement without filler is 34%, on the 7th – 42%, on the 28th – 55%.

The reason for the relatively low value of α is the extremely small amount of water required for the specimen without its squeezing (W/C ≈ 0.12). The composition consisting of 75% cement and 25% of granite dust ($m_{0.16}$) (particle size < 0.16 mm) was

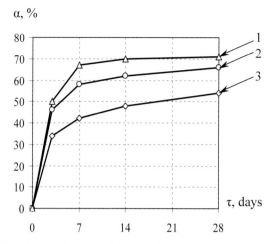

Fig. 4.5 Influence of granite dust ($m_{0.16}$) on vibro-pressed cement stone hydration degree: $1 - C = 50\%$, $m_{0.16} = 50\%$; $2 - C = 75\%$, $m_{0.16} = 25\%$; $3 - C = 100\%$.

formed at W/C = 0.15. At the same time, the value of α has increased significantly: on the 3rd day – 46%, 7th – 58% and 28th – 66%. Replacement of 50% cement by granite dust (W/C = 0.17) allowed an increase to the hydration degree at 3 days by 47% (α = 50%), 7 days by 59% (α = 67%) and 28 days by 32% (α = 71%), that is, to approximate those values observed for a cement paste of normal consistency [25]. Thus, adding granite dust significantly intensifies the hydration process of vibration pressed cement stone, which positively affects the VFC strength.

Table 4.2 The compositions of concrete specimens for investigating the kinetics of ultrasonic waves passing velocity vs. time.

No.	C, kg/, m³	A, kg/, m³	W, l/m³	$m_{0.16}$, %	W/C
1	313	1807	125	0	0.4
2	353	1730	141	9	0.4
3	395	1647	158	18	0.4
4	172	1955	115	0	0.67
5	190	1900	127	9	0.67
6	206	1864	138	18	0.67

Note: C, A and W are consumptions of cement, aggregate and water respectively.

4.3 Influence of granite siftings on water demand, output coefficient and strength of vibro-pressed concrete

To study the strength properties of VFC on granite siftings, series of experiments were conducted using three-factorial plan B_3. During the first series of experiments, the following factors varied: C/W (X_1), the content of particles < 0.16 mm in granite sifting $m_{0.16}$, (X_2), the content of granite sifting (m_s) in a mixture with ($M_f = 2.1$)

conditioned quartz sand (X_3). Terms of experiments planning are given in Table 4.3. The studied parameters were water demand (W), output factor (K_o), $K_o = \rho_r^n / \rho_b$ (ρ_r – actual average density of the mixture, kg/m^3; ρ_b – bulk density of the mixture before compacting), compressive strength (f_c) and flexural strength ($f_{c.\,tf}$) at 28 days.

Table 4.3 Experiments planning conditions for obtaining Eqs. (4.10 ... 4.13).

Factors		Variation levels			Variation interval
Natural	Coded	−1	0	+1	
C/W	X_1	1.4	2.2	3	0.8
Content of particles < 0.16 ($m_{0.16}$), %	X_2	0	9	18	9
Content of granite siftings in the mix with quartz sand, m_s, %	X_3	50	75	100	25

As a result of statistical processing of the experimental data, adequate confidence probabilities of 95% of the regression equation were obtained:

$$W = 109 + 19X_1 + 23X_2 + 23X_3 + 7X_1^2 + 6X_2^2 - X_3^2 - 3X_2X_3 \tag{4.10}$$

$$K_o = 0.54 - 0.01X_1 - 0.01X_2 - 0.05X_3 + 0.03X_3^2 + 0.01X_2X_3 \tag{4.11}$$

$$f_c = 31.8 + 11.9X_1 + 5.4X_2 - 0.6X_3 - 2.1X_2^2 - 0.4X_3^2 - 2.7X_1X_2 + X_1X_3 + 0.4X_2X_3 \tag{4.12}$$

$$f_{c.tf} = 6 + 1.5X_1 + 0.9X_2 - 0.3X_3 - 0.5X_1^2 - 0.5X_2^2 - 0.5X_3^2 - 0.4X_1X_2 + +0.2X_1X_3 + 0.1X_2X_3 \tag{4.13}$$

Analysis of the concrete mixture water demand equation shows that the influence of factor X_{1i} is of particular importance. The dependence of $W = f$ (C/W) has a progressive growth character, i.e., it is obvious that the well–known water demand constancy rule [65] practically does not apply to concrete mixtures compacted by vibro-pressing (Fig. 4.6).

With an increase of medium sized quartz sand in the aggregate content the concrete mixture water demand decreases in average from 130 to 82 l/m^3, i.e., 1.6 times (Fig. 4.7).

Factors X_1–X_3 also affect the output coefficient of the concrete mixture (Figs. 4.8, 4.9), with the highest influence of factor X_3 – the siftings content in mixture with sand. Adding up to 50% of quartz sand with M_f = 2.1 yields an increase in K_o from 0.52 to 0.6 ... 0.63. The most significant effect was observed at its content of 12 ... 50%. Other factors have little effect on K_o. As it was expected, from all varied factors, the VFC compressive and flexural strengths are most influenced by factor X_1 (C/W).

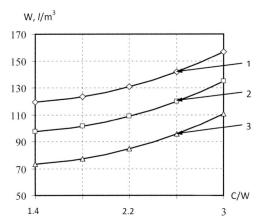

Fig. 4.6 Dependence of super stiff concrete mixtures water demand on the cement-water ratio: $1 - m_{0.16} = 18\%$; $2 - m_{0.16} = 9\%$, $3 - m_{0.16} = 0\%$.

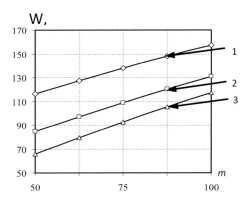

Fig. 4.7 Dependence of super stiff concrete mixtures water demand on the granite siftings content in aggregate: $1 - m_{0.16} = 0$; $2 - m_{0.16} = 9\%$; $3 - m_{0.16} = 18\%$.

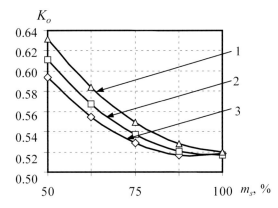

Fig. 4.8 Dependence of VFC output coefficient on granite siftings content in aggregate: $1 - m_{0.16} = 0$; $2 - m_{0.16} = 9\%$, $3 - m_{0.16} = 18\%$.

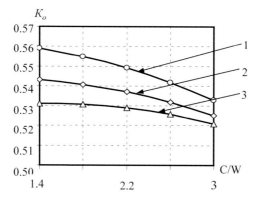

Fig. 4.9 Dependence of VFC output coefficient on cement – water ratio: $1 - m_{0.16} = 0$; $2 - m_{0.16} = 9\%$; $3 - m_{0.16} = 18\%$.

Transmission of C/W from bottom to upper level (Table 4.3) leads to an average increase in f_c from 20 to 42 MPa. It should be noted that the increase in C/W leads to a more significant increase in compressive strength than in the flexural one [13]. This effect is also known for other concrete types. With an increase in C/W f_c increases linearly: as the quadratic effect b_{22} in Eq. (4.14) is absent.

In contrast to the compressive strength, the growth of $f_{c.tf}$ with an increase in C/W to the maximum varied value has a nonlinear character (Figs. 4.10, 4.11).

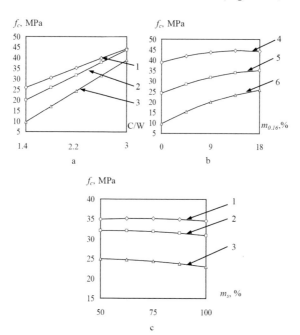

Fig. 4.10 Influence of C/W (a), content of particles < 0.16 mm ($m_{0.16}$) (b) and content of siftings in a mixture with quartz sand m_s (c) on VFC compressive strength: $1 - m_{0.16} = 18\%$; $2 - m_{0.16} = 9\%$; $3 - m_{0.16} = 0\%$; $4 - C/W = 3$; $5 - C/W = 2.2$; $6 - C/W = 1.4$.

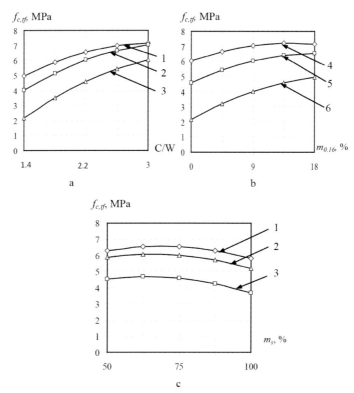

Fig. 4.11 Influence of C/W (a), particles < 0,16 $m_{0.16}$ (b) and content of siftings in a mix with quartz sand m_s (c) on VFC flexural strength (correcting additive – coarse sand): 1 – $m_{0.16}$ = 18%; 2 – $m_{0.16}$ = 9%; 3 – $m_{0.16}$ = 0%; 4 – C/W = 3; 5 – C/W = 2.2; 6 – C/W = 1.4.

At C/W, fixed at a certain level, an increase in the content of dusty impurities in granite siftings leads to VFC compressive and flexural strength growth. This conclusion is not consistent with the effect of dusty particles on the strength of conventional concrete at traditional technology, but is well-connected with theoretical ideas about the effect of fillers in concrete [13] with limited water demand. For vibro-pressed concrete with limited cement paste content, dust particles increase the cement matrix volume, which, to a certain extent, contributes to increasing the concrete density and strength, after which the influence of this factor becomes opposite.

The X_2 factor contributes to increase in flexural strength ($f_{c,tf}$) slightly more than in compressive strength (f_c). If increase from 0 to 18% in particle content < 0.16 mm yields an average increase of f_c by 37 ... 48%, $f_{c,tf}$ becomes higher by 42 ... 50%. Quadratic effect of factor X_3 indicates that the "efficiency threshold" of granite dust from the position of achieving maximum strength is within the range of 15 ... 22%. The smallest particles of granite taking an active part in the processes of hydration and the formation of the crystals nuclei, contribute to a more intense crystallization, which is reflected in the cohesive and adhesion capacity of the filled cement stone. Larger particles of fraction < 0.16 mm in turn increase the cement matrix volume, which causes improvement of concrete macrostructure, reduces the volume of cavities that are not filled with cement, improves the molding properties of super stiff concrete mixture, which leads to a decrease in the amount of air entrained at vibro-pressing. Confirmation of this kind of granite fillers influence is present in the regression equations of interaction coefficient for factors X_1 and X_2. Either at reducing the cement consumption, or at decrease in C/W, the positive effect due to increase in the content of particles < 0.16 mm significantly increases. If at C/W = 3 compressive strength increases by 13 ... 14%, then at C/W = 1.4–2.4 about 2.7 times. The highest impact of particles < 0.16 mm was found at insufficient content of cement paste to fill the aggregate voids at concrete vibro-pressing. At C/W = 2.5 ... 3, the optimal value of the factor X_2 is close to the main (zero) level (Table 4.3). The concrete strength begins to decrease when the filled cement paste spreads the aggregate grains, and the basic load is less taken by the aggregate stone skeleton, but more by the cement stone with filler. Adding to the aggregate conditioned quartz sand in addition to the siftings positively affects both the compressive and the flexural strength. Adding 50% of sand by the weight of siftings yields an average increase in concrete strength of 8 ... 10%. The positive effect degree of sand as part of aggregate increases with a decrease in C/W. At C/W = 3 (upper level, Table 4.3) factor X_3 becomes completely unimportant (Fig. 4.11). It is obvious that presence of conditioned quartz sand in aggregate reduces the overall aggregate voidness, therefore at low C/W, when the cement paste volume is insignificant to form a dense concrete structure, using such an additive is most effective. This assumption is confirmed by similar interaction of factors X_3 and X_2 in the strength Eqs. (4.12, 4.13).

In the second series of experiments, the following factors varied: X_1 - cement consumption (C = 300 ± 100 kg/m³), X_2 – content of particles < 0.16 mm in granite siftings ($m_{0.16}$ = 9 ± 9% of the siftings weight), X_3 is the content of granite siftings in a mixture with fine quartz sand (M_f = 1.1) (m_s = 75 ± 25%). The obtained and based on the experimental data of the regression equations for strength are given below:

$$f_c = 24.3 + 7.8X_1 + 4X_2 + 2.2X_3 + 2.7X_2^2 + \\ 2.4X_3^2 - 2.5X_1X_2 - 0.8X_1X_3 + 1.5X_2X_3 \tag{4.14}$$

$$f_{c,tf} = 4 + 1.1X_1 + X_2 + 0.6X_3 - 0.3X_1^2 - \\ 0.3X_2^2 + 0.3X_3^2 - 0.2X_1X_2 + 0.2X_1X_3 + \\ 0.4X_2X_3 \tag{4.15}$$

Analysis of model (4.14) shows that with an increase in cement consumption from 200 to 400 kg/m³ f_c increases 2.2 times (Fig. 4.12). The flexural strength $f_{c,ft}$ increases from 2.5 to 4.7 MPa (by 84 ... 95%). Either with constant C/W, or constant cement consumption the increase in the content of particles < 0.16 in granite siftings contributes to increasing the VFC strength, with the increase in cement consumption, the effectiveness of disperse granite filler is reduced. Fine sand with a considerable amount of dusty impurities has ambiguous effect on VFC properties, as is evident from the presence of a significant quadratic and interaction effects in the corresponding equation. On average, 25% of such sand has negligible effect on f_c and $f_{c,tf}$, 50% causes a decrease in f_c by 25% and $f_{c,tf}$ – by 30 ... 32%. The highest

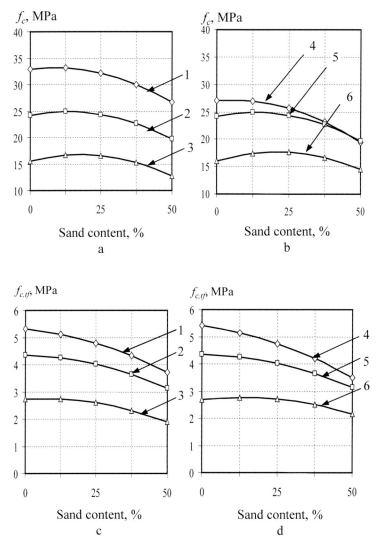

Fig. 4.12 Influence of fine sand content in aggregate on VFC compressive strength (a, b) and on flexural strength (c, d): $1 - C = 400$ kg/m^3; $2 - C = 300$ kg/m^3; $3 - C = 200$ kg/m^3; $4 - m_{0.16} = 18\%$; $5 - m_{0.16} = 9\%$; $6 - m_{0.16} = 0\%$.

increase in strength due to increase in sand content is observed when fixing other factors at the lower level. The positive effect of this component is possible only at a low binder amount. In this case, a lot of voids are formed in concrete, and can be reduced by using such an additive. Higher content of fine quartz sand with a significant amount of emulsified impurities causes a significant increase in the water demand of concrete mix and reduced strength.

4.4 Effect of the fractions volume ratio in siftings on VFC properties

Along with the influence of siftings' fraction < 0.16 mm grain composition, it is interesting to evaluate the effect of the larger fractions ratio. Investigation of granite siftings' grain composition effect on VFC properties was carried out according to B_3 plan [50]. As varied factors were taken: X_1 – volume concentration of fraction 10 ... 2.5 mm in a mixture of fractions 10 ... 2.5 mm and 2.5 ... 0.63 mm; X_2 – volumetric concentration of the mixture of fractions 10 ... 2.5 mm and 2.5 ... 0.63 mm in the total volume of washed granite siftings; X_3 – the fraction content < 0.16 mm. The experiments planning conditions are given in Table 4.4.

Table 4.4 Experiment planning conditions for obtaining Eqs. (4.19 ... 4.23).

Factors			Variation levels			Variation interval
Natural		Coded	−1	0	+1	
$\dfrac{V_{10...2.5}}{V_{10...2.5} + V_{2.5...0.63}}$		X_1	0.354	0.443	0.532	0.0886
$\dfrac{V_{10...2.5} + V_{2.5...0.63}}{V_{10...2.5} + V_{2.5...0.63} + V_{0.63...0.16}}$		X_2	0.618	0.773	0.928	0.155
Content of particles < 0.16 mm ($m_{0.16}$), %		X_3	0	10	20	10

Taking into account the planning conditions, the volume content of varying fractions at each point of the experimental plan was determined as follows:

$$V_{10...2.5} = X_1 \cdot X_2 \qquad (4.16)$$

$$V_{2.5...0.63} = (1 - X_1) \cdot X_2 \qquad (4.17)$$

$$V_{0.63...0.16} = 1 - X_2 \qquad (4.18)$$

The research was carried out using a concrete mixture composition: cement (C) - 400 kg/m³, aggregate (A) - 1650 kg/m³. During the experiments, the concrete mixture water demand, the average concrete density (ρ_0, kg/m³), the VFC output coefficient (K_o) and the concrete strength properties were controlled. Cylindrical specimens were prepared by the vibration method with a load of 0.06 MPa and the vibration duration was 20 seconds. The concrete mixture water demand was obtained by the condition of achieving the required specimens molding, which was controlled by the appearance of the liquid phase squeezing signs from the molded specimen. As a result of the experiment, regression equations were obtained. These equations are experimental-statistical models of the following studied (output) parameters:

$$W = 152 - 6X_1 - 3X_2 + 11X_3 - 12X_1^2 + 1X_2^2 - 6X_3^2 + 3X_1X_2 - 3X_1X_3 + 6X_2X_3 \qquad (4.19)$$

$$\rho_o = 2086 - 23X_1 - 5X_2 + 51X_3 - 22X_1^2 - 5X_2^2 -$$
$$25X_3^2 + 6X_1X_2 + 10X_1X_3 + 40X_2X_3 \qquad (4.20)$$

$$K_o = 0.52 + 0.08X_1 + 0.02X_2 - 0.01X_3 + 0.03X_1^2 +$$
$$0.01X_2^2 - 0.01X_1X_3 \qquad (4.21)$$

$$f_c = 29.9 - 4.4X_1 - 0.9X_2 + 4.1X_3 - 6.5X_1^2 +$$
$$3.1X_2^2 - X_3^2 + 1.6X_1X_2 + 2.6X_1X_3 + 3.3X_2X_3 \qquad (4.22)$$

$$f_{c.f} = 5.2 - 0.5X_1 - 0.2X_2 + 1.4X_3 - 0.4X_1^2 + 0.5X_2^2 -$$
$$0.4X_3^2 + 0.4X_1X_2 - 0.2X_1X_3 + 0.4X_2X_3 \qquad (4.23)$$

Analysis of granite siftings' grain composition influence makes it possible to state that the water demand of super-stiff concrete mixtures is uniquely linked to the efficiency modulus M_e [60] (Table 4.5):

$$M_e = \frac{\rho_a - \rho_{a.c.}}{\rho_a \rho_{a.c.}} + 0.013S \qquad (4.24)$$

where ρ_a and $\rho_{a.c}$ are density of the aggregate before and after compacting by pressing; S is the specific surface of the aggregate.

The specific surface of aggregate S can be approximately estimated [13] as

$$S = \frac{6.35k}{1000}(0.5a + b + 2c + 4d + 8e + 16f + 36g) \text{ m}^2/\text{kg} \qquad (4.25)$$

where k is a coefficient, depending on type of sand (for stone siftings and their mixtures with quartz sand $k = 2$); $a ... f$ are residuals on standard sieves with hole size from 5 to 0.16 mm, %; g - passage through a sieve with openings 0.16 mm, %.

The efficiency modulus characterizes the cement paste amount, in l, required to fill the voids and lubricate the grains in 1 kg of aggregate with a film of some conventional thickness (13 microns). At some constant cement consumption, the change in M_e when the aggregate grain composition changes, should be characterized by the corresponding change in water demand (Fig. 4.13).

Our results show that, although at various cement consumption the absolute water demand value for concrete mixtures changes, and the water demand depending on M_e remains practically the same.

The water demand dependence on M_e can be approximated by a logarithmic function (per 1 m³ of the mixture):

$$W = 4.53 \cdot ln\frac{M_e - 0.2}{0.005} + 59 \qquad (4.26)$$

Extrapolating the obtained concrete mixture water demand values beyond the varying area boundaries and conducting additional verification tests allowed the

Table 4.5 Relation between the water demand of the super-stiff concrete mixes and aggregate efficiency modulus.

Aggregate grain composition, %			M_e	Water demand, l/m^3		Deviation,%	Correlation coefficient
10–2.5	2.5–0.63	0.63–0.16		Experimental	Calculated by Eq. 4.26		$r_{y/x}$
49.4	43.4	7.2	0.206	116	118.3	1.94	0.976
32.9	28.9	38.1	0.221	128	128.8	0.62	
32.9	53.9	7.2	0.208	116	120.9	4.05	
21.9	39.9	38.1	0.253	137	137.3	0.22	
41.2	36.2	22.6	0.211	121	123.6	2.10	
27.4	50.0	22.6	0.218	126	127.9	1.49	
41.2	51.6	7.2	0.232	128	132.9	3.69	
27.4	34.,4	38.1	0.268	145	139.5	3.94	
34.3	43.1	22.6	0.244	135	135.7	0.52	

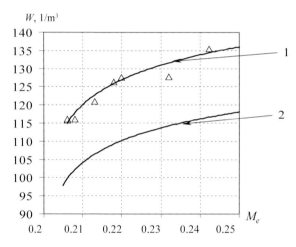

Fig. 4.13 Dependence of water demand on the concrete efficiency modulus $1 - C = 400$ kg/m³ $2 - C = 295$ kg/m³.

construction of triple diagrams for dependence of this. Additional experiments made it possible to construct triple diagrams of the water demand dependence for a concrete mixture on the granitic sifting (Fig. 4.14).

Of all the factors that characterize the grain composition of granite siftings, washed from dust impurities, the highest effect on the water demand has factor X_1. An increase in fraction 10 ... 2.5 mm leads to a sharp decrease in water demand, especially in absence or low content of fraction 2.5 ... 0.63 mm (up to 20%). Further increase in fraction 2.5 ... 0.63 mm and increase in the content of a large fraction (10 ... 2.5 mm) to 40% leads to some increase in W.

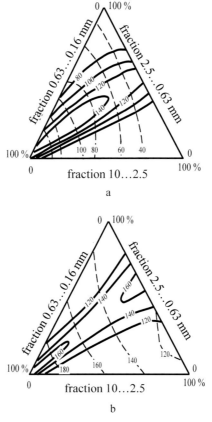

Fig. 4.14 Dependence of concrete mixture water demand
(W, l/m^3, ——) and specific surface of granite siftings (S, sm/g- - - -) on filler grain composition: (a) washed granite siftings; (b) content of particles < 0.16 mm 20%.

The highest effect on water demand of the investigated concrete mixture creates an increase in factor X_3 – the number of particles < 0.16 mm. However, it should be noted also some influence of other fractions content (especially 0.63 ... 0.16 mm and 10 ... 2.5 mm) with the increase in the dust particles content.

Varying the granite siftings grain composition is strongly reflected in the change of the vibro-pressed concrete average density. Thus, transition of factor X_1 from the lower to the upper level while fixing other factors on the main one leads to a decrease of ρ_0 from 2195 to 2040 kg/m³ (Fig. 4.15). This demonstrates that increasing content of fraction 10 ... 2.5 mm by 20% from the initial value is the reason for the increase in the number of cavities in vibro-pressed concrete, which are not filled with cement paste. Increase in content of particles < 0.16 mm (X_3) reduces the negative effect of the large fraction in such a way that factor X_1 becomes completely unimportant.

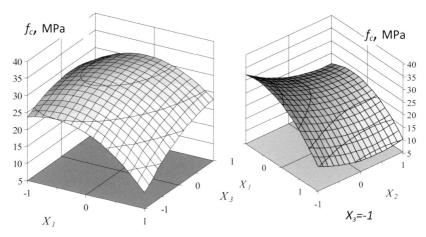

Fig. 4.15 Influence of granite siftings grain composition on VFC compressive strength (X_1, X_2 X_3 - investigated factors (Table 4.4)).

Transition of factor X_1 from the lower to the upper level contributes to the increase of the VFC output coefficient, since increasing the large fraction content (10 ... 2.5 mm) in a mixture of fractions of 10 ... 2.5 mm and 2.5 ... 0.63 mm leads to an increase in the produced vibro-pressed concrete volume. An increase in the content of fine fraction (0.63 ... 0.16 mm) also causes an increase in the output coefficient, but to a lower extent. With increasing a large fraction content, dust particles increase the compacting ability of the mixture on the granite siftings.

Increasing the content of 10 ... 2.5 mm fraction within a varying region leads to a decrease in f_c, although the W/C decreases. This fact is caused by an increase in the aggregate voidness with an increase in the content of large particles. At the same time, the constant cement consumption during the study leads to the fact that the cement paste becomes insufficient to fill the voids and the formation of a binding agent film on the aggregate grains. With an increase in the content of particles < 0.16 mm (factor X_3) influence of factor X_1 on strength: becomes extreme. The strength optimum is observed with the content of a large fraction of 30 ... 40% at dust amount of 15 ... 18%. This confirms our preliminary conclusions that particles < 0.16 mm contribute to increasing VFC strength and density, increasing the cement paste volume is not enough to fill the voids.

Comparing the values of the VFC compressive and flexural strength, it is noticeable that their ratio for this concrete type is within the range of 4 to 5, when it is known that for ordinary concrete $f_c/f_{c.tf} = 8 ... 5$ [61]. Decreasing this parameter to a certain extent characterizes the improvement of composite material structure. Production of FGC from the super-stiff concrete mixes by vibro-pressing contributes to a more ordered structure formation. At the same time, a positive role in this process plays filler, which is the disperse granite particles (Fig. 4.16).

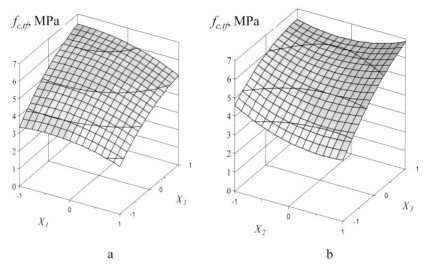

Fig. 4.16 Influence of granite siftings grain composition on VFC flexural strength: (X_1, X_2, X_3 - investigated factors (Table 4.4)).

4.5 Influence of plasticizers and air entrained admixtures

Reducing of super-stiff concrete mixtures viscosity at constant water demand allows better mixture compacting and removing the excess air. As a result, concrete porosity decreases and its strength increases. Plasticizing admixtures, which are widely used in plastic and cast concrete mixtures, yield a slight plasticizing effect in stiff mixtures [60]. A reason for this is the relatively small amount of liquid phase in such mixtures, which prevents surface-active substances to fully realize the plasticization effect.

As is well known [42], the highest effect from using plasticizers in plastic concrete mixtures is obtained as a result of reducing their water demand, and with constant cement consumption - due to lower W/C of concrete. This enables to obtain concrete with higher strength at C = const, or to save cement when achieving a certain strength. For these reasons it was found the minimum water demand, required to achieve the suitable molding requirements for stiff concrete mixtures at adding naphthalene-formaldegide superplasticizer (SP) The experimental results are given in Table 4.6 and in Fig. 4.17. At cement consumption C = 400 kg/m³, the concrete mixture water demand decreased by 11 *l*/m³, at C = 200 kg/m³, only 3 *l*/m³ (Fig. 4.17).

When performing the second series of experiments at tests with the same cement consumption, water demand of concrete mixes remained constant (Fig. 4.18).

At constant water demand, adding SP admixture leads to a significant increase in concrete average density (by 170 ... 180 kg/m³) (Fig. 4.18).

Application in the super-stiff concrete mixtures air entraining admixtures (AEA) is attractive. The admixture content varied in the range from 0 to 0.1% of the cement weight, the results of experiments are given in Table 4.6 and in Fig. 4.19.

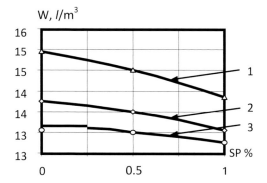

Fig. 4.17 Effect of superplasticizer (SP) on water demand of super stiff concrete mixtures 1 – C = 400 kg/m³; 2 – C = 300 kg/m³; 3 – C = 200 kg/m³.

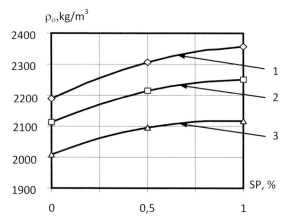

Fig. 4.18 Effect of superplasticizer SP on average density of concrete mixes after vibro-pressing (W = const): 1 – C = 400 kg/m³; 2 – C = 300 kg/m³; 3 – C = 200 kg/m³.

Analyzing the obtained data, it should be noted that the effect of AEA admixture on ρ_0 has a sharply expressed extreme nature – with an increase in the admixture content to the optimum (0.06 ... 0.075%), the average density of the vibro-pressed concrete mixture increases from 2080 to 2170 kg/m³

It is well known [66] that when adding air entraining admixtures into plastic concrete mixtures composition, due to hydrophobization of the solid phase surface there is entraining at mixing small air bubbles, leading to some plasticization of the concrete mixture and a decrease in its average density. The result that was obtained in our study, obviously, is caused by the fact that the insignificant plasticization action of air-entraining admixture, which is evident in plastic concrete mixtures, becomes dominant in super-stiff ones. Air bubbles, involved into the concrete mix at mixing, act as microcells, which reduce friction between the compacted mixture particles and contribute to their more compact positioning. At the same time, some air redistribution in VFC is probable, as a result of which a part of the remaining after compacting air that is harmful to concrete passes into emulsified state, which positively affects the

Table 4.6 SAS admixtures affect on vibro-pressed concrete properties.

Concrete composition	Additive content, %	SP								AEA			
		Method of adding								Additive content, %	ρ_0, kg/m³	W_o, %	f_c, MPa
		Without changing the concrete mixture water demand				With changing the concrete mixture water demand							
		C/W	ρ_0, kg/m³	W_o, %	f_c, MPa	C/W	W, l/m³	ρ_0, kg/m³	f_c, MPa				
C = 400 kg/m³ A = 1650 kg/m³	0	2.56	2189	12.3	31.8	2.56	156	2210	32.2	0	2169	11.7	31.9
	0.5	2.56	2306	11	39.3	2.67	150	2232	35	0.05	2236	9.2	35.7
	1	2.56	2358	10.3	42.9	2,8	143	2263	37.4	0.1	2208	9.8	32.4
C = 300 kg/m³ A = 1720 kg/m³	0	2.1	2112	13.6	28.8	2.1	143	2139	29.2	0	2086	13.1	27.3
	0.5	2.1	2214	12.7	35	2.14	140	2159	31	0.05	2168	10.6	32.4
	1	2.1	2250	12.4	37.3	2.22	135	2188	32.6	0.1	2155	11.2	30.1
C = 200 kg/m³ A = 1780 kg/m³	0	1.48	2010	17	19.6	1.48	135	2010	20.2	0	1968	16.2	22.9
	0.5	1.48	2096	16.4	23.8	1.48	135	2028	21.1	0.05	2065	13.8	29.4
	1	1.48	2116	16.5	24.8	1.52	132	2056	21.8	0.1	2067	14.4	28.4

ρ_0 – average concrete mix density after vibro-pressing, kg/m³; W_o – volumetric water absorption C – cement content; A – aggregate content.

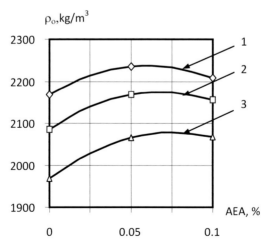

Fig. 4.19 Effect of AEA admixture on average density of compacted concrete: 1 – C = 400 kg/m³; 2 – C = 300 kg/m³; 3 – C = 200 kg/m³.

frost resistance. With a decrease the cement paste amount in the concrete mix the optimal air entraining admixture content increases.

Analysis of the obtained results allows us to assert that the adding of air entraining admixture in an optimum content results increase in strength by an average of 20% (Fig. 4.20). The negative effect of interaction between variable factors indicates a certain increase in the optimum AEA content in concrete with a reduced cement paste amount. The optimal content of this admixture varies from 0.068 to 0.051%.

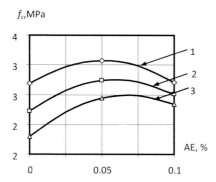

Fig. 4.20 Influence of air entraining admixture on VFC strength:
1 – C = 400 kg/m³; 2 – C = 300 kg/m³; 3 – C = 200 kg/m³.

Investigation of the joint effect of SP and AEA admixtures was carried out using the experimental three-level plan B₂ for two factors. At all the experiments, the concrete composition was constant: C = 400 kg/m³, A = 1650 kg/m³, W = 157 l/m³. As aggregate (A) were used granite siftings containing dusty particles $m_{0.16}$ = 18%. Adequate at 95% probability experimental-statistical models for average density of compacted concrete mixture (ρ_0, kg/m³) and VFC compressive strength at 28 days (f_c, MPa) are given below:

$$\rho_0 = 2351 + 58X_1 + 28X_2 - 19X_1^2 - 42X_2^2 \tag{4.27}$$

$$f_c = 41.5 + 4.7X_1 + 1.8X_2 - 2.1X_1^2 - 0.9X_1X_2 \tag{4.28}$$

The equations are valid in the range of the following factors values: SP content (X_1) 0.5 ± 0.5%, AEA (X_2) 0.03 ± 0.03% of cement weight. The combined effect of the investigated admixtures causes an additional increase in average density of the concrete mixture – the factors interaction coefficient in Eq. (4.27) is absent. If increase in SP admixture content causes an increase in ρ_0 from 2194 to 2319 kg/m³ and for AEA an increase in ρ_0 is up to 2267 kg/m³, mixing these admixtures in optimal ratio (SP-1%, AEA-0.045%) allows an increase in ρ_0 up to 2393 kg/m³. The amount of residual after compacting air in VFC at a combined addition of plasticizing and air-entraining admixtures in optimal amount was reduced from 11.3 to 2.8%. Combined action of SP and AEA causes a 41% increase in strength (30 to 43 MPa). Due to adding

AEA admixture, the content of SP can be significantly reduced without reducing the VFC strength. Effective composition of the complex admixture can be determined using the iso-parametric diagram (Fig. 4.21), which is based on model (4.28).

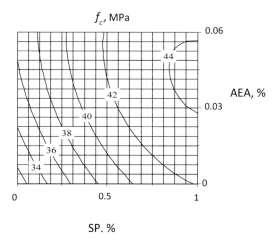

Fig. 4.21 Combined affect of SAS admixtures on VFC strength.

4.6 Effect of heat treatment on VFC properties

One of the most used methods in the industrial technology of concrete products is their thermal treatment by steam. The least sensitive to increased temperature is concrete, made of stiff concrete mixtures that undergo heat treatment in densely closed molds or under pressure, i.e., in conditions that prevent their expansion at hardening. Provisions on heat treatment recommend to set the temperature raising rate during steaming in accordance with the initial concrete strength, i.e., strength before heating [13]. For concrete, for which this characteristic is more than 0.6 MPa, the temperature rise rate that avoids significant negative consequences, can be taken 50 ... 60°C/hour. Based on these data, it can be assumed that when steaming VFC from super-stiff concrete mixtures even forced temperature rise modes are permissible, since its strength immediately after molding is twice higher than the recommended by the standards. However, it should be considered that when vibro-pressing a concrete mix, up to 100 l/m^3 of residual air remains in concrete. As known, this air creates a negative effect during heat treatment of concrete. As a result of thermal expansion, all the concrete components increase in volume, but the highest linear thermal expansion coefficients have water and air, which increase in volume 2 ... 3 times, and cause an internal pressure, sometimes reaching 0.01 ... 0.015 MPa that can significantly worsen the concrete structure [67].

In order to obtain the optimal heat treatment mode of VFC on granite siftings, a factor experiment was implemented according to B_4 plan. The following factors were varied: X_1 - isothermal heating temperature (t_{is} = 60 ± 20°C), X_2 - isothermal heating duration (τ_{is} = 7 ± 2 h), X_3 - cement-water ratio (C/W = 2 ± 0.5), X_4 - content of particles < 0.16 mm in granite siftings ($m_{0.16}$ = 9 ± 9%).

The specimens were made by the method of vibration with loading of 0.06 MPa and molding duration of 20 sec. Steaming was carried out in a laboratory steaming chamber at the following mode: $2 + 2 + \tau_{is} + 2$. As studied parameter was used compressive strength at 4 hours after heat treatment (f_c, MPa).

After statistical analysis of experimental data, the regression dependence of concrete compressive strength on the above selected factors is obtained in coded variables:

$$f_c = 11.7 + 1.7X_1 - 0.3X_2 + 2.2X_3 + 3.6X_4 + 0.8X_1^2 -$$
$$2.1X_2^2 + 0.3X_3^2 - 0.9X_4^2 - 0.2X_1X_3 + 0.6X_1X_4 + \qquad (4.29)$$
$$0.3X_2X_3 + 0.6X_2X_4 + 0.6X_3X_4$$

According to Eq. (4.29), the factors can be placed in the following sequence by their influence on the strength after steaming:

$$X_4(m_{0.16}) > X_3(C/W) > X_1(t_{is}) > X_2(\tau_{is}).$$

The presence of particles < 0.16 mm in granite siftings positively affects the steamed VFC strength. When steaming, adding granite filler is effective also at high values of C/W.

Increase in the isothermal heating temperature (X_1) unambiguously leads to an increase in the strength of the concrete. Presence of a positive quadratic effect in the equation indicates the strength growth at the transition of this factor from the lower variation level to the upper. Attention is drawn to the positive mutual influence coefficient of factors X_1 and X_4 - heating temperature and content of filler particles. A more significant increase in strength with a simultaneous increase in temperature and filler content, on the one hand, can be explained by the intensification of the filled cement paste structuring processes at a higher temperature, and on the other – by influence of the residual air (which the content in concrete without filler particles < 0.16 was 1.5 ... 2 times more) at higher temperature rise rate.

The lowest linear effect on concrete strength value has factor X_2 - isothermal heating duration, but the presence in Eq. (4.29) of corresponding significant quadratic effect with negative sign indicates the extreme nature of the effect of isothermal heating duration on strength in the selected variation region.

Some authors [68, 69] noted a strength decrease with increasing the isothermal heating duration instead of the expected growth. This is explained by the fact that at a certain stage of concrete isothermal heating its original structure changes, there are internal stresses that contribute to defects formation and decrease the cement stone strength. The ongoing hydration process leads to the disappearance of cracks and defects under the influence of new hydration products. The optimal duration of isothermal concrete treatment increases with increasing C/W and granite filler content and is within the range of 6.5 ... 7.5 hours (Fig. 2.22).

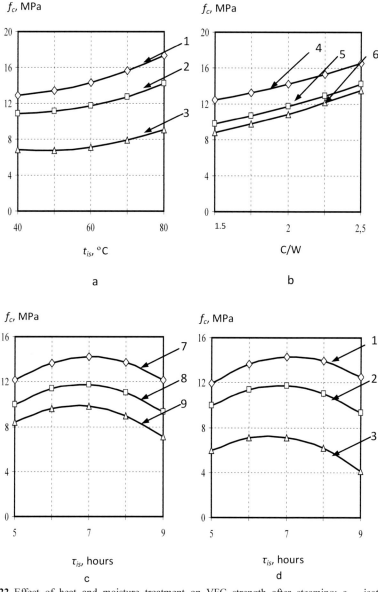

Fig. 4.22 Effect of heat and moisture treatment on VFC strength after steaming: a – isothermal heating temperature; b – cement-water ratio; c, d – isothermal heating duration; $1 – m_{0.16}$ = 18%; $2 – m_{0.16}$ = 9%; $3 – m_{0.16}$ = 0%; $4 – t_{is}$ = 80°C; $5 – t_{is}$ = 60°C; $6 – t_{is}$ = 40°C; $7 – $ C/W = 2.5; $8 – $ C/W = 2.0; $9 – $ C/W = 1.5.

Steaming of the VFC on granite siftings at 80°C and optimal duration gives an opportunity to get at 4 hours after heat treatment 63 ... 67% of 28–days concrete strength.

It is known that from the viewpoint of lowering the thermal energy cost and improving the concrete strength, combined use of heat treatment and hardening accelerators admixtures is effective [66]. To study the possibility of reducing the of isothermal heating temperature, experiments were conducted with two accelerating admixtures – calcium chloride (CC) and sodium thiosulfate (ST) on strength of VFC with granite siftings. A three-level plan was applied for three factors B_3 [50]; X_1 (C/W = 2 ± 0.5); X_2 (t_{is} = 60 ± 20°C); X_3 (accelerator admixture 1 ± 1%). The heat treatment mode was 2 + 2 + 7 + 2, its duration τ_{is} was based on the average of its optimum values, obtained from Eq. (4.28).

Adequate regression equations for strength of steamed concrete with acceleration admixture are given below:

- at adding CC:

$$f_c = 16.1 + 2.3X_1 + 1.7X_2 + 1.8X_3 + 0.3X_1^2 + +1.2X_2^2 - 1.4X_3^2 + \\ 0.4X_1X_2 + 0.2X_1X_3 + 0.2X_2X_3 \tag{4.30}$$

- at adding ST:

$$f_c = 17.9 + 2.4X_1 + 1.8X_2 + 1.4X_3 + 0.4X_1^2 + X_2^2 - 2.8X_3^2 + \\ +0.5X_1X_2 + 0.2X_1X_3 + 0.1X_2X_3 \tag{4.31}$$

According to the coefficients of Eqs. (4.30, 4.31), in general the effect of varying factors on f_c is similar, despite the use of various admixtures.

To study the effectiveness of these admixtures for reducing the isothermal heating temperature, isoparametric analysis of the obtained equations was carried out. The isolines f_c in coordinates D (admixtures content), % and t_{is}, °C are presented in Fig. 4.23. The results of analysis indicate that in this variation region the maximum strength during steaming, which is 24 ... 25 MPa and is 74 ... 76% of the design value, is obtained by adding admixtures in optimal quantities and maximum steaming temperature of 80°C.

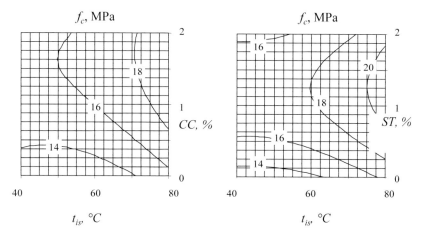

Fig. 4.23 Effect of accelerating admixtures: a – calcium chloride; b – ST on VFC stress after heat and moisture treatment.

Adding CC and ST admixtures at optimal concentrations reduces the steaming temperature to 40 ... 450°C without significant strength reduction.

To find optimal admixtures' concentrations Eqs. (4.30, 4.31) were differentiated by X_3 and as a result the optimum equations in natural form were obtained:

- adding ST(D_1^{opt}):

$$D_1^{opt} = 0.08\,C/W + 0.00125ts + 0.7 \qquad (4.32)$$

- adding CC(D_2^{opt}):

$$D_2^{opt} = 0.14\,C/W + 0.0045t_{is} + 1.1 \qquad (4.33)$$

It can be summarized that an increase in steaming temperature and especially C/W leads to an increase in the admixtures concentration, at which the highest effect of strength increasing after steaming is achieved. The optimal admixtures concentrations are given in Table 4.7.

Table 4.7 Optimal concentrations of accelerating admixtures.

Admixture type	Isothermal heating temperature, °C	Admixture content, % of cement weight		
		C/W = 1.5	C/W = 2.0	C/W = 2.5
Calcium chloride (CC)	40	1.51	1.58	1.65
	80	1.69	1.76	1.82
Sodium thiosulfate(ST)	40	0.94	0.98	1.01
	80	0.98	1.02	1.06

4.7 Structure and properties of VFC determining its durability

Significant influence on concrete structures has hydraulic shrinkage, which becomes evident at concrete drying. It is due to the action of capillary forces that arise in cement stone during water evaporation from capillaries, removal of inter-crystal and adsorption-bound water from tobemorite gel [13]. Shrinkage deformation (ε_{sh}) causes internal stresses in concrete, which cause cracks in the contact zone and contributes to lower frost resistance and impermeability.

Shrinkage deformations of VFC are significantly lower than for the usual concrete due to the almost complete lack of free and capillary moisture in them. The obtained results (Table 4.8) indicate that the presence of granite filler in VFC does not increase the value of ε_{sh} and in some cases even reduces it. This is consistent with available data showing that when adding fillers, there is a decrease of concrete shrinkage due to adsorption bonding of water and as a result of the cement stone structure ordering and technological damage reduction [13].

Adding superplasticizer reduces the concrete shrinkage (Table 4.8) due to the reduction of the cement paste layers thickness between the aggregate grains and facilitating a more dense concrete structure formation. Concrete resistance to the destructive effects of various kinds is largely determined by the structure of their porous space [63].

As a result of experiments according to B_3 plan [50], regression equations for total and open (imaginary) porosity (P, % and P_o, %, correspondingly) were obtained for 3, 7 and 28 days, as well as for the mean pore size (λ) and pore size homogeneity (α). The open porosity was estimated on the basis of volumetric water absorption of specimens.

Table 4.8 Deformation properties of VFC.

No.	Concrete composition, kg/m³			W/C	$m_{0.16}$, % f_c, MPa	Content of SP, %	Experimental data			
							Shrinkage deformation $\varepsilon \sim 10^5$			
	C	A	W				28 days	90 days	120 days	180 days
1	400	1650	161	0.4	20	-	32.3	5.4	6.7	6.8	7.0
2	400	1650	125	0.31	0		23.4	7.0	8.9	9.2	9.5
3	400	1650*	116	0.29	-		35.0	4.8	5.8	6.0	6.2
4	200	1780	135	0.63	20		17.8	2.8	3.6	3.7	3.9
5	200	1780	123	0.61	0		6.4	4.6	5.5	5.7	6.0
6	200	1780*	100	0.5	-		21.0	2.3	3.2	3.4	3.5
7	400	1650	161	0.4	20	1	45.8	3.7	4.8	4.9	5.1
8	200	1780	135	0.63	20		24.0	1.8	2.6	2.8	2.9
9	400	1650	160	0.4	20	-	37.6	5.3	6.3	6.4	6.5
10	400	1650	160	0.4	20	1	39.1	3.3	4.3	4.5	4.7

Notes: 1. The specimens were formed by vibration with loading of 0.055 MPa 2. In series marked with asterisks, quartz sand was used as filler ($M_f = 2.0$), in other series granite siftings were used.

The equations have the following form:

$$y = b_0 + \sum b_i x_i + \sum b_{ii} x_i^2 + \sum b_{is} x_i x_s \qquad (4.34)$$

The regression equations' coefficients are given in Table 4.9.

The equations are valid in the following variable factors range: cement consumption C (X_1) = 300 ± 100 kg/m³; content of particles < 0.16 mm in granite siftings $m_{0.16}$ (X_2) = 9 ± 9%; content of granite siftings in a mixture with quartz sand ($M_f = 1.1$) (X_3) = 75 ± 25%.

At 28 days, the total porosity of the studied VFC was in the range of 14.5 ... 26%, the open porosity −10 ... 24%. The factor contributing to the highest reduction of both total (P) and open porosity (P_o) is cement consumption (X_1): transition of X_1 from the lower to the upper level causes a relative change of P by 75%, and P_o by 40 ... 45% (Fig. 4.24). It significantly reduces porosity of VFC increasing the content of particles

< 0.16 mm in granite siftings (X_2): the total porosity was increased by 22% and open by 38 ... 40%. Reducing porosity in this case is more significant at low values of C/W, and consequently lower cement consumption, when the cement paste is not sufficient to form a dense concrete structure. Comparing the obtained mathematical models of VFC porosity at different hardening periods, it should be noted that the presence of particles < 0.16 mm in granite siftings affects both the magnitude of P and P_0 and the kinetics of their changing in time.

Table 4.9 Coefficients of regression equations.

Properties	Age, days	b_0	b_1	b_2	b_3	b_{11}	b_{22}	b_{33}	b_{12}	b_{13}	b_{23}
General porosity P, %	3	18.6	−2.4	−1.6	−0.1	1.3	0.8	0.1	0.25	-	−0.73
	7	16.7	−2.53	−1.68	−0.113	1.32	0.72	0.12	0.32	−0.04	−0.99
	28	15.6	−2.67	−1.79	0.13	1.36	0.76	0.10	−0.31	−0.05	−1.00
Open porosity P_o, %	3	15.4	−2.52	−2.41	−0.04	1.31	1.19	−1.60	0.55	−0.24	−1.52
	7	13.4	2.48	−2.50	−0.05	1.28	1.25	−1.61	0.58	−0.21	−1.65
	28	12.5	−2.54	−2.53	−0.05	1.32	1.22	−1.56	0.61	−0.22	−1.62
λ	28	2.28	−0.19	−0.13	−0.01	0.06	−0.01	−0.02	0.09	−0.02	−0.02
α	28	0.46	0.11	0.03	0.01	0.02	−0.01	−0.02	−0.02	−0.02	0.01

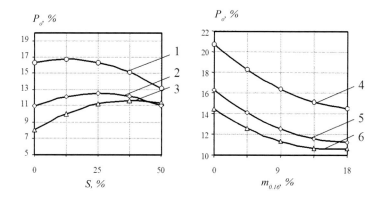

Fig. 4.24 Dependence of open porosity of VFC on cement consumption (C), content of corrective quartz sand additive (S) (a) and content of particles < 0.16 mm ($m_{0.16}$) in granite siftings (b): $1 - m_{0.16} = 0\%$; $2 - m_{0.16} = 9\%$; $3 - m_{0.16} = 18\%$; $4 - C = 200$ kg/m³; $5 - C = 300$ kg/m³; $6 - C = 400$ kg/m³.

Reduction of concrete porosity during hardening occurs mainly during the first 7 days: at $m_{0.16} = 20\%$, total porosity decreases in average by 12%, and the open one by 17%; at $m_{0.16} = 0\%$ - P decreases by 10.5%, and P_o - by 8.5%. Obviously, particles < 0.16 mm in granite siftings accelerate the rate of decrease in total VFC porosity, increase the proportion of conditionally closed porosity in the total pores volume.

Among the variable factors the highest influence on the parameters λ and α have by X_1 and X_2, i.e, the cement consumption and the content of particles < 0.16 mm in granite siftings. With transition of these factors from the lower to the upper level, there is a decrease in the average pore size and increase in their uniformity. This is especially noticeable in concrete with low cement consumption, in which, due to insufficient cavities filling by cement paste, the largest and most varied in size pores are observed. An increase in C/W from 1.5 to 2.5 causes an appropriate decrease of λ from 2.3 to 2.0 and an increase in α from 0.42 ... 0.61 (Fig. 4.25).

Particles < 0.16 mm affect porosity to a lesser extent: λ decreases on average from 2.12 to 2.0, and α rises from 0.55 to 0.61. Analysis of experimental data (Table 4.9) shows that adding superplasticizer to concrete mixtures contributes to reducing the water absorption volume and accordingly open porosity. On average, 1% of SP by cement consumption reduces P_o by 40%. Reducing C/W of concrete causes a decrease in this effect, as indicated by the negative interaction coefficient of these factors in the equation.

A different nature on the open porosity effect has air entraining admixture (Fig. 4.26). Up to some limit in AEA content (0.06–0.07%) P_o decreases and after this a value increase is observed. The reason for the increase in P_o may be an increase in the open pores proportion, caused by under-compacting of the concrete mixture. The optimal admixture amount contributes to a decrease in P_o by 33%.

Along with the individual influence of plasticizers and air entraining admixtures on the open porosity of VFC (P_o), their complex influence was evaluated. The regression equation for P_o obtained from the experiment carried out using the factor plan B_2 is shown below:

$$P_o = 9,22 - 1,3X_1 - 0,9X_2 + 0,63X_1^2 + 0,6X_2^2 + 0,57X_1X_2 \qquad (4.35)$$

The concrete composition at the admixtures content variation remained constant - C = 400 kg/m³, granite siftings - 1650 kg/m³ and water - 157 l/m³. The equation is

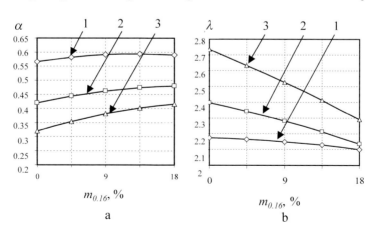

Fig. 4.25 Effect of granite filler on pore size uniformity indicators, α (a) and average pores' size, λ (b) of VFC: 1 – C = 400 kg/m³; 2 – C = 300 kg/m³; 3 – C = 200 kg/m³.

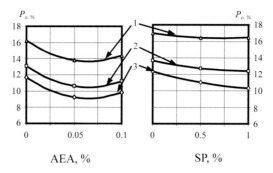

Fig. 4.26 SAS admixture affect on open porosity of VFC: 1 – C = 200 kg/m³; 2 – C = 300 kg/m³; 3 – C = 400 kg/m³.

valid in the following range of the factors' values: SP (X_1) - 0.5 ± 0.5%; AEA (X_2) – 0.03 ± 0.03%. The presence of a significant interaction coefficient of these factors indicates that the nature of the individual influence of admixtures changes with their complex addition (Fig. 4.26). If superplasticizer causes a decrease in specimens water absorption by 47 ... 48%, combined effect of SP and AEA is 68% (P_o = 7.2%).

Concrete porosity significantly affects its frost resistance. As is known [49], destruction of concrete during alternating freezing and thawing in a saturated water state occurs mainly due to the tensile forces resulting from the increase in the volume of ice thus formed. Concrete ability to withstand fracture during repeated freezing and thawing is due to the presence of reserve pores filled with air in its structure. Part of water is squeezed into these pores under the action of ice crystals growing at freezing.

As a frost resistance criterion in experimental studies it was used as a coefficient of frost resistance K_f, which was determined as the ratio of compressive strength of cylindrical samples $d = h = 50$ mm after a certain number of cycles of repeated freezing

Table 4.10 Concrete compositions for determining VFC frost resistance.

No.	Concrete composition, kg/m³			Content of particles < 0.16 mm in aggregate, %	Additives availability, quantity (%)
	Cement	Aggregate	Water		
1	400	1650	157	18	-
2	400	1650	125	-	-
3	400	1650*	116	-	-
4	200	1780	125	18	-
5	200	1780	123	-	-
6	200	1780*	100	-	-
7	400	1650	157	18	SP, 1
8	400	1650	157	18	AEA, 0.06
9	400	1650	157	18	SP, 1;a, 0.03

*in these concrete compositions as aggregate was used quartz sand (M_f = 1.9), in other compositions – granite siftings.

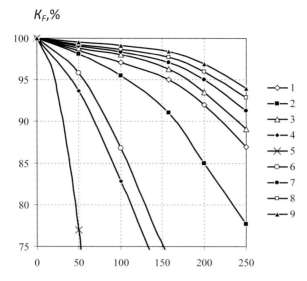

Fig. 4.27 Frost resistance of VFC.
Note: positions in the figure correspond to the order in Table 4.10.

and thawing and strength of control samples at the corresponding age. Concrete compositions are given in Table 4.10.

Analyzing the obtained experimental data, it can be concluded that the tested VFC has a frost resistance at a standard strength reduction of 5% in the range of 50 to 230 cycles. Frost resistance is significantly increased with a decrease in concrete W/C. On average, reduction of W/C from 0.7 to 0.38 leads increased frost resistance from 50 to 150 cycles (Fig. 4.27). Adding plasticizers, in particular, complex plasticizing-air-entraining admixtures significantly increases VFC frost resistance.

The dominant factors of influence on concrete water impermeability, corrosion and frost resistance is peculiarities of concrete porous structure and first of all the through pores volume and their dimensions. In the case of vibro-pressing concrete products from super-stiff concrete mixtures, sedimentation processes, which are the main source of through pores formation in plastic concrete mixtures, practically do not occur. Due to low concrete mixture water demand, when compacting by vibro-pressing a significant amount of water is not squeezed, therefore directed micro-capillaries are not common for the formed concrete structure. The main factors contributing to the water absorption reduction and the increase of the VFC frost resistance should also be positively reflected in their impermeability and corrosion resistance. The results of concrete samples' impermeability tests are given in Table 4.11. The obtained data are in good agreement with results of the open porosity study.

For products made of vibro-pressed concrete, working in aggressive environment, sulfate resistance is an important property. This feature was determined by changing the splitting tensile strength after the samples were kept in 5% Na_2SO_4 solution for 360 days. The results of the tests are shown in Table 4.11.

Table 4.11 Water impermeability and sulfate resistance of VFC.

Composition No. according to Table 4.10	Splitting tensile strength ($f_{t.s.}$, MPa) in 5% Na$_2$SO$_4$ solution after, days			Water impermeability (MPa) after, days		
	28	180	360	28	180	360
1	3.14	3.18	3.21	0.47	1.16	1.41
2	2.54	2.46	2.36	0.11	0.28	0.34
3	3.32	3.42	3.38	0.68	1.66	2.02
4	2.11	2.01	1.86	0.03	0.08	0.10
5	1.07	0.87	0.78	0.01	0.02	0.02
6	2.36	2.31	2.15	0.07	0.17	0.21
7	3.97	4.29	4.44	1.04	2.56	3.11
8	3.44	3.61	3.79	0.87	2.12	2.58
9	3.98	4.39	4.50	1.18	2.89	3.52

As expected, samples without chemical admixtures had the lowest resistance to the investigated corrosion type. Adding SP and AEA significantly increase the VFC corrosion resistance due to concrete porosity reduction. A slightly lower positive effect was obtained by adding into the VFC filler a form of granite siftings particles < 0.16 mm.

4.8 Composition design of VFC with stone crushing siftings

Composition design methodology of VFC on granite siftings with particles < 16 mm acting as filler is based on the obtained empirical dependencies.

The main components of VFC on granite siftings are related by a system of linear equations:

$$\begin{cases} V_c + V_W + V_f + V_a + V_{\text{air}} = 1000 \\ V_c + V_W + V_f = M_e A \end{cases} \tag{4.36}$$

where V_c, V_W, V_f, V_a, V_{air} are absolute volumes of cement, water, filler, aggregate and air, respectively, l; A is the aggregate content, kg/m^3; M_e – efficiency modulus.

From Eqs. (4.36) the contents of aggregate (A) and filler (m_f) are obtained:

$$A = \frac{1000 - V_{air}}{M_e - \dfrac{1}{\rho_a}} \tag{4.37}$$

$$m_f = \left(1000 - V_{air} - \frac{m_a}{\rho_a} - \frac{C}{\rho_c} - W \right) \cdot \rho_f \tag{4.38}$$

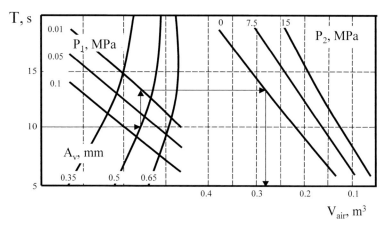

Fig. 4.28 Nomogram for determining the residual air volume in VFC: T – compacting duration, sec.; A_v – vibration amplitude, mm; P_1 – load at vibration, MPa; P_2 – compaction pressure, MPa; (vibration frequency is 50 Hz).

where ρ_c, ρ_a, ρ_f are densities of cement, aggregate and filler, respectively, kg/l; C and W are cement consumption and water demand, kg, V_a is the aggregate volume, V_{air} is the air volume.

The air volume significantly depends on the method and parameters of the concrete mixture compacting. Based on the obtained dependences, a model describing the compacting parameters effect on the residual air amount was compiled and a corresponding nomogram was constructed (Fig. 4.28).

To calculate the concrete C/W, which should provide the specified compressive strength, the following equation can be applied:

$$f_c^{28} = AR_c(C/W - b) \qquad (4.39)$$

Values of coefficients A and b for VFC on granite siftings, calculated based on mathematical models for strength are given in Table 4.12.

VFC compositions can be calculated either for molded mixture volume ($V_{m.m}$), or for the pressed concrete volume (V_{VFC}). These values are inter-related by output

Table 4.12 Values of coefficients A and b for VFC on granite siftings.

Content of siftings in a mix with coarse quartz sand (m_s), %	Content of granite filler, % of aggregate weight			
	20	15	10	0
100	A = 0.38 b = 0.1	A = 0.38 b = 0.25	A = 0.37 b = 0.39	A = 0.36 b = 0.7
75	A = 0.36 b = −0.08	A = 0.35 b = 0.06	A = 0.34 b = 0.2	A = 0.33 b = 0.5

coefficient (K_o), which is obtained as a ratio of the concrete mixture average density in loose-bulky state to that after compacting

$$V_{VFC} = V_{m.m} \cdot K_o \qquad (4.40)$$

The VFC output coefficient depends mainly on the characteristics of the filler used. Experimental values of K_o for concrete mixes with filler represented by granite siftings only is 0.51–0.52, at adding 25% of coarse sand it is 0.53 ... 0.54, and for 50% of sand – 0.6 ... 0.62.

Water demand per 1 m^3 of molded concrete mixture can be calculated using the following equation:

$$W_{m.m.} = 5.46\,(C/W)^2 + 0.068m_f^2 + 0.0021m_s^2 - 23C/W + 0.17m_f - $$
$$0.12m_s + 0.32C/W \cdot m_f + 0.096C/W \cdot m_s - 0.01m_f m_s + 54{,}66 \qquad (4.41)$$

Equation obtained by translating Eq. (4.10) in natural form, considering the data from Table 4.3 and using the following equation:

$$X_i = X_{io} + x_i \cdot \Delta X_i \qquad (4.42)$$

where X_i is the value of the factor in natural form, X_{io} is the value of the factor at the main (zero) level, x_i is the value of the factor in coded formi, ΔX_i is the factor variation interval.

As Eq. (4.40) is valid for granite siftings with $M_e = 0.223$, the mixture water demand should be corrected considering the M_e of each individual aggregate according to the following approach:

$$W = W_{m.m} + \Delta W$$

$$\Delta W = 4.53\ln\left(\frac{M_e - 0.2}{0.005}\right) - 6.91 \qquad (4.43)$$

In Eqs. (4.41, 4.43) the water demand is obtained per 1 m^3 of concrete mixture. Calculating it per 1 m^3 of concrete should be carried out considering K_o.

Composition design of VFC on granite siftings includes the following stages:

1. Taking into account the concrete strength, cement strength and aggregate features, estimate C/W from Eq. (4.40).
2. Obtain the approximate water demand ($W_{m.m.}^1$) and the corrected value (W^1) from Eqs. (4.41).
3. Considering K_o find the water demand per 1 m^3 of VFC ($W_c^1 = W^1/K_o$).
4. Using condition C = C/W·W, find the approximate cement consumption (C^1).
5. From the nomogram (Fig. 4.29), taking $m_f = 9\%$ and considering the compacting parameters, the content of residual air V_{air}^1 is estimated.
6. From Eqs. (4.39, 4.40) the approximate content of filler m_a^1 and granite microfiller m_f^1 are obtained.

7. Knowing the required content of filler, the values of C/W, water demand (W_c), cement consumption (C), content of residual air (V_{air}), aggregate content (m_a) and filler content (m_f) are specified.

Example. Design VFC compositions with 28-days compressive strength 20 and 30 MPa using Portland Cement with standard compressive strength 40 MPa and granite siftings with $M_e = 0.216$, dusty particles content of 18% without sand. Compacting parameters: $A_v = 0.5$ mm, P = 0.08 MPa, T = 15 s.

The design of VFC compositions is given in Table 4.13.

Table 4.13 Design of VFC compositions.

Concrete strength, MPa	Estimated values								
	C/W	$W^1_{m.m.}$, l	W^1, l	W^1_c, l/m^3	C^1, kg/m^3	V^1_{air}, % (l)	m^1_a, kg/m^3	m^1_f	
								kg/m^3	%
20	1.74	58.65	57.02	109.7	191.2	16.62 (166.2)	1415	360	20,2
30	2.76	72.49	70.86	136.3	376	12.7 (127)	1481.8	167	10.1

Concrete strength, MPa	Specified values							
	C/W	$W_{m.m.}$, l	W, l	V_{air}, % (l)	Concrete composition			
					W_c, l/m^3	C, kg/m^3	m_a, kg/m^3	m_f
								kg/m^3	%
20	1.42	75.8	74.17	11.27 (112.7)	142	203	1506	315.5	17.3
30	2.76	72.3	70.67	12.72 (127.2)	135.9	374	1481	170.6	10.3

CHAPTER 5

Macroporous Fine Grained Concrete Based on Stone Crushing Waste

5.1 General features of macroporous concrete technology (MPC)

The idea of producing macroporous concrete (MPC) that has a structure characterized by a considerable volume of intergranular cavities, was first proposed in 1912 by N.A. Zhitkevich [70]. Subsequently, the technology of this light concrete was studied by many researchers [71]. To obtain concrete, both light porous aggregates and ordinary heavy materials like gravel or crushed stone are used. Along with other types of light concrete, MPC can be used as a material for monolithic and precast wall constructions, as well as for drainage systems and filters. When using a heavy aggregate, the thickness of the walls made of MPC is equal to brick ones, while using porous aggregates, the wall thickness decreases 1.5 ... 2 times.

Macroporous concrete is a non-sand material that is usually obtained from a mixture of dense or porous gravel or crushed stone, cement and water with a limited content of cement paste.

The feature of MPC is its unique structure with a higher volume of cavities between grains, unlike in classical concrete.

The density of MPC on heavy aggregate usually does not exceed 1800 kg/m³, which allows its effective application for products and structures with high strength at a relatively low density. The porous structure of MPC allows its application also for products and structures with high filtration capability.

Selection of MPC composition can be carried out in any way, allowing production of material with specified strength, frost resistance and density of mixture with necessary workability at minimum cement consumption.

The size of aggregates grains for concrete is usually within the range of 5 ... 40 mm. Using more homogeneous and relatively fine aggregate reduces the MPC density and thermal conductivity coefficient as well as its air permeability [71].

Using mono-fraction aggregates (for example, 10 ... 20 mm), increases the binder consumption by 10%.

It is recommended to use graphic dependencies to determine the approximate W/C value of macro-porous concrete on a certain type of aggregate (Fig. 5.1). The actual W/C values are usually between 0.45 and 0.55.

To increase the MPC mixtures fluidity, surface-active additives, as well as lime in a form of finely milled powder or lime paste are used. To accelerate the concrete mixture hardening salts are used separately or in combination with surface-active additives. Consumption of cement, aggregate and water specify the test mixtures for preparing the control cubic specimens. At the same time, the selected composition should ensure the MPC mixture's non-dissipation. A concrete mix with properly selected composition is characterized by:

• homogeneity and uniform envelopment of aggregate grains by cement paste;
• absence of cement paste trickling from aggregate grains during concrete casting;
• non-dissipation of the concrete mixture during its transportation and at casting.

When producing MPC, the dosage accuracy (by weight or volume if required) is: for cement, additives and water ± 1% and for aggregates ± 2%.

Mixing of the concrete components is recommended to be carried out:

• when using dense aggregates in forced mixers and in concrete mixers of gravitational action;
• for light aggregates - only in mixers of forced action.

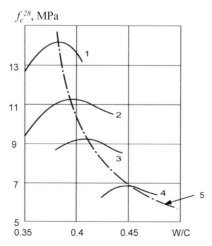

Fig. 5.1 Dependence of macroporous concrete strength on W/C: 1 – concrete composition (cement: gravel by volume) 1:6; 2 – concrete composition (cement: gravel by volume) 1:7; 3 – concrete composition (cement: gravel by volume) 1:8; 4 – concrete composition (cement: gravel by volume) 1:10; 5 – optimum curve of W/C.

Concrete mix molding is done by layers of 20 ... 30 cm with a uniform compacting of each layer by vibration on vibroplatform or external vibrators. The vibration duration, as a rule, should not exceed 10 ... 15s, in order avoid cement paste drainage from the aggregate surface.

One of the original technologies of MPC production was proposed by S.M. Izkovich [71]. Following this method, the quantity of the prepared cement paste is more than is required according to calculations. Then it is mixed with the aggregate. The obtained mass is subjected to short-term treatment on vibration sieve, leaving the remnant of the cement paste to be separated and returned for re-use, and the remainders of the concrete mixture are used to produce MPC. The cement paste quantity in such concrete equals to that which can keep the aggregate on its surface.

5.2 Properties of macroporous concrete based on stone crushing siftings

The purpose of the research was to investigate the influence of composition factors on the main properties (strength and average density) of MPS manufactured by vibro-pressing method. Using as aggregate pure granite siftings with mono-fractional composition enables us to assume that it is possible to produce concrete with optimal combination of strength and average density.

To produce MPS the following method was implemented. Preliminary pure granite siftings (fractions 2.5 ... 5 mm), Portland cement, water and additives (superplasticizer SP) were mixed. Then, after thorough mixing during 1 ... 1.5 min, the concrete mixture was put into a mold (cubic specimens $10 \times 10 \times 10$ cm) and placed on the vibration stand. The concrete mixture was molded in 3 layers with uniform compacting of each layer under loading of 0.3 N/cm^2. The specimens (Fig. 5.2) hardened 28 days at 20 ... 25°C and relative humidity of about 95%.

In order to obtain quantitative dependencies, taking into account the influence of main technological factors on the MPC properties, experiments have been carried out using mathematical planning. The main factors that varied during the experiments were cement consumption (X_1); W/C (X_2), content of additive SP (X_3).

Fig. 5.2 Laboratory specimen of MPC on granite siftings.

Table 5.1 presents the conditions for planning the experiment. The output parameters were taken, the concrete strength determined after 28 days of normal hardening (f_c^{28}) and average concrete density ρ_c.

Table 5.1 Experiment planning conditions for receiving Eqs. 5.1 and 5.2.

Factors	Variation levels			Variation interval
	−1	**0**	**+1**	
Cement consumption (X_1), kg	200	240	280	40
W/C, (X_2)	0.36	0.38	0.4	0.02
SP content (X_3)	0	0.3	0.6	0.3

Experimental values of 28-days strength and concrete mixture density, obtained by implementing a three-level three-factor plan are presented in Table 5.2.

Table 5.2 Experimental results.

No.	W/C	Components consumptions, kg/m³				Compressive strength, MPa (28 days) f_c^{28}	Average density, kg/m³ ρ_c
		C	**W**	**SP**	**Siftings (fraction 2–5 mm)**		
1	0.4	280	112	1.68	1600	16.5	1805
2	0.4	280	112	0	1600	12.9	1766
3	0.36	280	100.8	1.68	1600	11.8	1798
4	0.36	280	100.8	0	1600	5.9	1774
5	0.4	200	80	1.2	1600	6.8	1679
6	0.4	200	80	0	1600	6.2	1717
7	0.36	200	72	1.2	1600	7.6	1722
8	0.36	200	72	0	1600	5.1	1689
9	0.38	280	106.4	0.84	1600	16.3	1810
10	0.38	200	76	0.6	1600	13.2	1716
11	0.4	240	96	0.72	1600	13.1	1792
12	0.36	240	86.4	0.72	1600	12.3	1770
13	0.38	240	91.2	1.44	1600	9.8	1789
14	0.38	240	91.2	0	1600	9.9	1759
15	0.38	240	91.2	0.72	1600	14.2	1770
16	0.38	240	91,2	0.72	1600	14.4	1772
17	0.38	240	91.2	0.72	1600	13.8	1789

After statistical processing the following experimental-statistical (mathematical) models for macroporous concrete strength (f_c^{28}) and average density (ρ_c) were obtained:

$$f_c^{28} = 14.1 + 2.31X_1 + 1.14X_2 + 1.39X_3 + 0.57X_1^2 - 1.48X_2^2$$
$$- 4.3X_3^2 + 1.25X_1X_2 + 0.98X_1X_3 - 0.35X_2X_3 \tag{5.1}$$

$$\rho_c = 1786 + 4X_1 + 8.8X_3 - 23.8X_1^2 - 4.5X_2^2 - 12.8X_3^2$$
$$+ 1.8X_1X_2 + 8.5X_1X_3 - 7X_2X_3 \tag{5.2}$$

To study the influence of varying factors on the concrete strength at the corresponding age, graphs of their influence were constructed (Fig. 5.3). The third factor was fixed at the main (zero) level (Table 5.1). Graphical dependences, obtained using these models, are shown in Figs. 5.3 and 5.4.

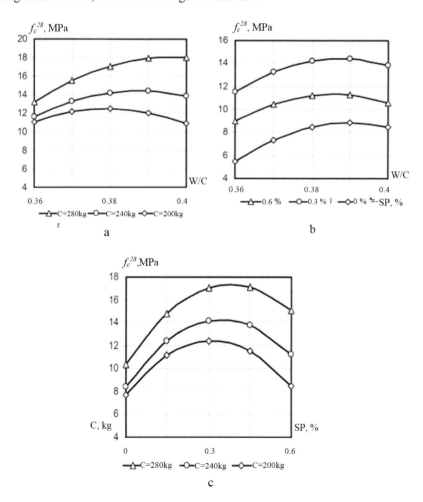

Fig. 5.3 Dependence of MPC compressive strength on technological factors: a – cement consumption and W/C; b – W/C and superplasticizer content; c – cement consumption and superplasticizer content.

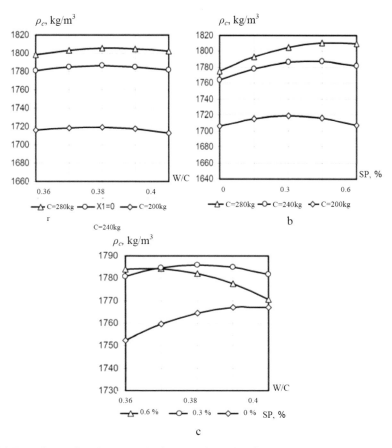

Fig. 5.4 Dependence of MPC average density on technological factors: a – cement consumption and W/C; b – cement consumption and superplasticizer content; c – W/C and superplasticizer content.

Analyzing mathematical models and graphic dependences on their basis, it can be concluded that the MPC strength depends on the cement stone strength and quantity. However, increasing the cement consumption does not always lead to adequate strengthening of contact between the aggregate grains. With decrease in W/C the influence of cement consumption on the MPC strength significantly decreases (Fig. 5.3).

The MPC strength is also affected by addition of superplasticizer. As follows from the models analysis (Fig. 5.3), there is a certain optimum of the superplasticizer content from the viewpoint of strength of this concrete type.

Experimental results show that the change in the W/C has an ambiguous effect on concrete strength. Increase in strength with increasing W/C and constant cement consumption to a certain extent is explained by the need to ensure a minimum water demand which creates conditions for binding aggregate grains in a single conglomerate. A dominant influence on MPC density has cement consumption (Fig. 5.4) and, accordingly, the cement stone quantity. Other variables have insignificant effect on the MPC density.

Thus, in view of the obtained results, it can be noted that with moderate cement consumption (240 ... 280 kg/m³) and SP additive (0.3 ... 0.6%), it is possible to obtain macro-porous concrete based on granites siftings fraction 2 ... 5 mm with the following characteristics: $f_c^{28} = 12 ... 16$ MPa; $\rho_c = 1750...1800$ kg/m³.

One of the effective ways for using MPC can be its application as an envelope for producing of so-called wall "Thermo-blocks" (Fig. 5.5).

Comparison of thermo-physical and physical-mechanical properties of such products with other wall materials is given in Table 5.3 and in Fig. 5.6.

Thus, macro-porous concrete obtained using fraction of 2.5 ... 5 mm of granite siftings is quite competitive compared to "classic" wall materials. Its thermal resistance is similar to ceramic and silicate bricks, but at the same time, substantially exceeding the specified materials in strength and, consequently, constructive quality coefficient – ratio of compressive strength and average density.

Fig. 5.5 External view of thermo-block envelope for thermo-block made of MPC and thermo-block with thermo-isolating infill from cellular concrete with a density of 500 kg/m³.

Table 5.3 Physical-mechanical and thermo-physical properties of wall products.

Compared materials	Density		Thermal conductivity, W/m·K	Thermal resistance, m²·K/W	f_c, MPa	Constructive quality coefficient, MPa·g/cm³
	kg/m³	g/cm³				
Ceramic brick (full)	1700	1.7	0.78	0.63	7.5	4.4
Silicate brick	1800	1.8	0.83	0.55	10	5.6
Foam concrete (D800)	800	0.8	0.31	2.17	3,2	4
Macro-porous concrete	1800	1.8	0.83	0.6	15	8.3
Macro-porous concrete	1700	1.7	0.78	0.64	12	7.1
Thermo-block	1160	1.16	0.492	1.017	10	8.6

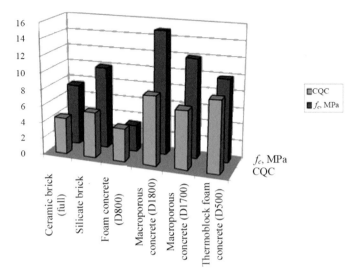

Fig. 5.6 Compressive strength (MPa) and constructive quality coefficient (CQC) of the compared materials.

CHAPTER 6

Normal-weight Concrete Based on Stone Siftings

6.1 Influence of granite siftings on normal-weight concrete properties

One of the possible applications of stone siftings is using them as a fine aggregate for normal-weight concrete. To determine the effectiveness of using granite siftings as fine aggregate of concrete, instead of natural sand, experiments were carried out for three factors according to B_3 plan [50] (experimental planning conditions are given in Table 6.1).

The concrete mixture fluidity at all plan points was maintained within the limits of the cone slump (CS) = 15 ... 18 cm. The given fluidity was achieved by adding a naphthalene-formaldehyde type superplasticizer (SP). For the obtained concrete mixture stratification was determined. Compressive strength of $10 \times 10 \times 10$ mm cubic specimens was determined at 7 and 28 days. The results of experiments are shown in Table 6.2.

As a result of the experimental data statistical analysis, adequate regression equations for the required superplasticizer amount,% to provide the specified concrete

Table 6.1 Experiments' planning conditions for obtaining Eqs. (6.1 … 6.4).

No.	Factors			Variation levels			Variation interval
	Natural	Coded		−1	0	1	
1	Water-cement ratio (W/C)	X_1		0.35	0.4	0.45	0.05
2	Part of siftings (n_s) ,% in the fine aggregate mass	X_2		0	50	100	50
3	Part of the fine aggregate in the aggregates' mixture (r), %	X_3		30	45	60	15

mixture fluidity (*SP*), concrete mixture stratification (St) and concrete strength at 7 (f_c^7) and 28 days (f_c^{28}) (MPa) are obtained:

$$SP = 0.62 - 0.04X_1 + 0.15X_2 + 0.06X_3 + 0.037X_1^2 + 0.187X_2^2$$
$$0.037X_3^2 + 0.075X_1X_2 - 0.025X_2X_3 \tag{6.1}$$

$$St = 2.69 - 0.79X_1 + 0.56X_2 - 1.16X_3 + 0.68X_1^2 - 0.46X_2^2 +$$
$$+ 0.22X_3^2 + 0.19X_1X_2 + 0.053X_1X_3 - 0.31X_2X_3 \tag{6.2}$$

$$f_c^7 = 38.74 - 2.48X_1 - 1.39X_2 - 5.18X_3 + 3.28X_1^2 - 0.6X_2^2$$
$$0.6X_3^2 + 4.2X_1X_2 + 0.11X_1X_3 + 1.06X_2X_3 \tag{6.3}$$

$$f_c^{28} = 41.34 - 11.41X_1 - 2.1X_2 - 3.9X_3 + 6.22X_1^2 - 2.34X_2^2 +$$
$$+ 0.18X_3^2 + 2.22X_1X_2 + 1.9X_1X_3 + 2.67X_2X_3 \tag{6.4}$$

The change in concrete mixture water demand was evaluated indirectly by Eq. (6.1), which reflects the influence of the investigated factors on the superplasticizer amount, providing the specified concrete mixture fluidity. The maximum increase in water demand, as it was expected, is caused by the factor X_2 (replacement of

Table 6.2 Results of experimental studies on properties of normal-weight concrete based on stone siftings.

Plan points	Natural values of factors			SP content, %	f_c^7 days, MPa	f_c^{28} days, MPa	St, %
	W/C	n_s, %	r, %				
1	0.45	100	60	1.2	39.6	49	2.98
2	0.45	100	30	1	46.9	53.5	4.66
3	0.45	0	60	0.6	29.8	44.7	1.0
4	0.45	0	30	0.6	42.6	47.8	3.69
5	0.35	100	60	1.1	30	32.7	4.18
6	0.35	100	30	1	47.5	54.0	5.86
7	0.35	0	60	1	46.8	55.3	2.74
8	0.35	0	30	0.7	50.3	56.7	5.86
9	0.45	50	45	0.6	38.8	50.5	3.22
10	0.35	50	45	0.6	47.9	53.9	4.89
11	0.4	100	45	0.8	35.2	42.8	4.44
12	0.4	0	45	0.7	43.7	48.5	3.24
13	0.4	50	60	0.7	35.3	43.9	2.4
14	0.4	50	30	0.5	46	52.4	4.8
15	0.4	50	45	0.7	36.8	46.3	3.84
16	0.4	50	45	0.7	36.8	46.3	0.7
17	0.4	50	45	0.7	36.8	46.3	0.7

conditioned sand by granite siftings), due to the increased content of particles < 0.16 mm in the siftings. The required superplasticizer content also increases significantly due to an increase in the proportion of sand in a mix of aggregates (factor X_3) - more than 30 ... 40%. The results of the research are presented in Figs. 6.1 ... 6.3.

The increase of W/C (factor X_1) causes a slight increase in the concrete mixture water demand, which is noticeable only outside the constancy of water demand rule [65]. The increased voidness and significant content of the needle grains in the siftings is the reason for the concrete mixture stratification which was observed (Eq. 6.1). Other factors (X_1 and X_3) contribute to significant reduction of stratification and increase the concrete mixture homogeneity. Stratification, caused by high content of siftings (factor X_2) is compensated by interaction with other factors. With increase in the part of sand in the aggregates' mix and reduction of W/C, the concrete mixture heterogeneity, which was caused by the presence of siftings, becomes less evident due to the increase in the volume of cement-sandy mortar in the concrete composition.

The concrete strength at 28 days was within 32 ... 55 MPa that basically corresponds to the strength of concrete in the investigated water-cement ratio range (X_1). Other factors also cause changes in the concrete strength, and by their effect it is possible to place them as follows: the maximum change in strength (25 ... 28%) is caused by factor X_3 (sand fraction), next - (10 ... 15%) by factor X_2 (the part of siftings in mixture with sand). For all factors in the strength models (Eqs. 6.3, 6.4), there is a positive interaction coefficient, indicating a decrease in the negative impact on strength while increasing or decreasing the values of the corresponding variables (Figs. 6.1, 6.2). The negative effect of the siftings content on strength is noticeable in the case when the fine aggregate part is minimal. In this case, the aggregate has higher voidness, which causes a decrease in strength. This coincides with the maximum

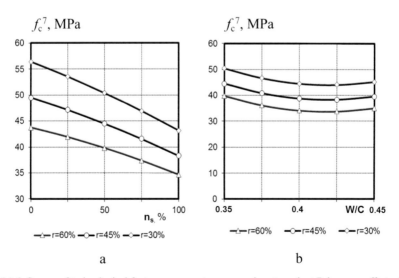

a   b

Fig. 6.1 Influence of technological factors on concrete compressive strength at 7 days: a – effect of fine aggregate part in the aggregates' mix; b – effect of water-cement ratio and fine aggregate part in the aggregates' mix.

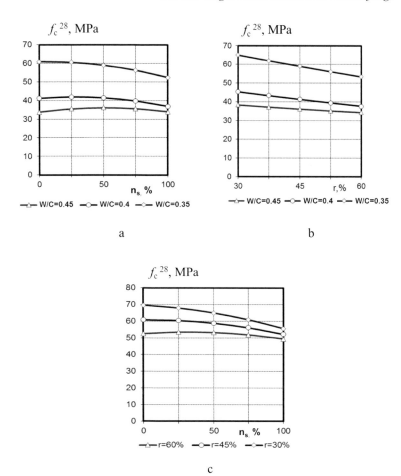

Fig. 6.2 Influence of technological factors on concrete compressive strength at 28 days: a – effect of water-cement ratio and part of siftings; b – effect of water-cement ratio and part of fine aggregate in the mix of aggregates; c – effect of siftings part and fine aggregate part in the aggregates' mix.

concrete mixture stratification. Increase in the fine aggregate part (r) enables us to use significantly more siftings without reducing the strength: at $r = 45\%$ - 35 ... 45%, at $r = 60\%$ - up to 70%. In this case a concrete mixture does not have stratification features, which is confirmed by analysis of the corresponding model.

6.2 Design of concrete compositions using stone crushing siftings

Design of concrete compositions with siftings can be carried out according to a well-known technique [61]. The content of natural sand, which should be added to the sand from the siftings, taking into account the part of particles > 0.63 mm is:

$$n = \frac{A_S - A_R}{A_S - A_N} \tag{6.5}$$

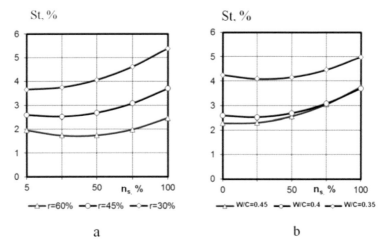

Fig. 6.3 Influence of technological factors on concrete mixture stratification: a – effect of siftings part and fine aggregate part in the aggregates' mix; b – effect of water-cement ratio and part of siftings.

where A_R, A_N, A_S are complete residues on sieve 0.63 mm of required, natural sands and sand from siftings, respectively.

The recommended grains content of siftings can vary within the limits given in Table 6.3.

The content of sand from the siftings (S_S) is obtained taking into account its ratio with natural sand (n):

$$S_S = S \cdot (1-n) \tag{6.6}$$

where S is the content of mixed fine aggregate in concrete, kg/m³.

The content of natural sand:

$$S_N = S - S_S \tag{6.7}$$

The change in the mixed fine aggregate density (ρ_A) due to the use of sand from siftings is taken into account by the following formula:

$$\rho_A = n \cdot \rho_{S_{Sn}} + (1-n)\rho_{S_s}, \tag{6.8}$$

where ρ_{Ss} and ρ_{Ssn} are densities of sand from siftings and natural sand, respectively, kg/m³.

Concrete mixes with high fluidity should be produced with obligatory application of superplasticizers. Chemical plasticizing admixtures should compensate the increase in the concrete mixture water demand caused by high content of dusty impurities in siftings.

Table 6.3 Recommended grain composition of siftings.

No.	Sieve openings, mm	Complete residue on control sieves, %
1	10	0 … 5
2	5	0 … 15
3	2.5	0 … 20
4	1.25	5 … 50
5	0.63	20 … 70
6	0.315	35 … 90
7	0.16	80 … 100

Example. *Determine the composition of concrete class C12/15 for a monolithic foundation. The concrete mix fluidity is cone slump 10 … 12 cm. Maximum crushed stone coarseness is 20 cm. As fine aggregate, is used sand from siftings (fineness modulus $M_f = 3.3$, the residue on sieve 0.63 mm is 71%, the content of particles < 0.16 mm is 16%). To adjust the grain composition of sand from siftings is used natural sand ($M_f = 1.9$, the residue on sieve 0.63 mm is 16%). To reduce the concrete mix water content is used superplasticizer SP.*

Characteristics of raw materials:

- Portland cement: with standard compressive strength $R_c = 52$ MPa, density of cement 3100 kg/m³, normal consistence of cement paste 26%;
- Sand from siftings: density $\rho_{S_s} = 2700$ kg/m³;
- Natural sand: real density $\rho_{S_i} = 2650$ kg/m³;
- Crushed stone fraction 5 … 20 mm: density $\rho_{C.S} = 2700$ kg/m³, bulk density $\rho_{bC.S} = 1350$ kg/m³; voidness $V_v = 0.47$.

1. By Eq. (6.5) find the required ratio between sand from siftings and natural sand, *n*, for the necessary 45% complete residue on sieve 0.63 mm:

$$n = \frac{A_S - 45}{A_S - A_N} = \frac{71 - 45}{71 - 16} = 0.47$$

Find the density of mixed sand by Eq. (6.8):

$$\rho_S = n \cdot \rho_{S_{Sn}} + (1 - n)\rho_{S_s} = 0.47 \cdot 2700 + (1 - 0.47)2650 = 2674 \text{ kg/m}^3$$

2. The water demand of the obtained mixed sand is found experimentally by the method proposed in [31]: $W_S = 9\%$ (higher than water demand of sand with average fineness by 2%). The concrete mixture water content should be increased by 10 *l* (2 × 5 *l*) on account of the aggregate water demand. The use of superplasticizer reduces the concrete mixture water demand by 20 liters.

3. Find the required W/C of the concrete mixture to provide the design strength (f_c = 20 MPa). Coefficient A in expression $f_c = AR_c \left(\dfrac{C}{W} - 0.5 \right)$ is equal to 0.55.

$$\dfrac{W}{C} = \dfrac{AR_c}{f_c + 0.5\,AR_c} = \dfrac{0.55 \cdot 52}{20 + 0.5 \cdot 0.55 \cdot 52} = 0.81$$

4. According to available data, find the water demand, considering correction related to water demands of sand, cement and superplasticizer:

$$W = 205 \; l/m^3$$

5. Cement consumption:

$$C = \dfrac{W}{\dfrac{W}{C}} = \dfrac{205}{0.81} = 253 \; kg/m^3$$

6. Content of crushed stone (C.S.):

$$C.S = \dfrac{1000}{\dfrac{1000}{\rho_{C.S}} + \alpha \cdot \dfrac{1000 \cdot V_v}{\rho_{bC.S}}} = \dfrac{1000}{\dfrac{1000}{2700} + 1.28 \cdot \dfrac{1000 \cdot 0.47}{1440}} = 1273 \; kg/m^3$$

where α is the coefficient of grains extension by cement-sand mortar. $\alpha = 1.28$.

7. The content of mixed sand is:

$$S = \left[1000 - \left(\dfrac{C}{\rho_C} + \dfrac{C.S}{\rho_{C.S}} + W \right) \right] \cdot \rho_S = \left[1000 - \left(\dfrac{253}{3.1} + \dfrac{1273}{2.7} + 205 \right) \right] \times 2.675$$

$$= 647 \; kg/m^3$$

of sand from siftings:

$$S_s = S(1 - n) = 647 \cdot (1 - 0.47) = 343 \; kg/m^3$$

8. Natural sand content:

$$S_n = S - S_s = 647 - 343 = 304 \; kg/m^3$$

Laboratory verification of the obtained concrete composition.

Experimental investigation has demonstrated the possibility of using granite siftings in the normal-weight concrete instead of fine aggregate, providing the optimal grain composition and allowing production of mixtures without stratification signs and the compensation of increasing the concrete mixture water demand by using superplasticizers.

CHAPTER 7

Dry Construction Mixtures and Mortars on their Basis Using Aspiration Dust

7.1 Aspiration granite dust as a possible disperse filler for mortars

Aspiration systems at crushing and sorting plants and equipment are set to control the dust formed during crushing and to clean the air before it is emitted in the atmosphere. Dust is captured in cyclones and at a two-stage purification also by bag filters.

One of the main areas of aspiration dusts possible application is using it as a disperse filler in various purpose dry construction mixtures [72].

Mineral fillers are used to reduce the cement consumption, improve the number of properties of mortars, including those based on dry construction mixtures. The content of fillers in dry construction mixtures can vary from 5 to 80%. Increasing the total content of binder due to fillers increases the mortar mixtures homogeneity, reduces their segregation, improves adhesion and other properties.

Known construction mixtures compositions include various types of powdered and fibrous fillers. The choice of the filler type and content should take into account the specific features of its effect on the construction mixtures properties. For example, the use of mica increases the cracking resistance of mortars, and also improves their thixotropic properties. Mineral fibers increase the mortars strength in bending and tension, as well as provide higher elasticity [73].

Fillers are characterized by mineral-petrographic and chemical composition, reactivity, bulk and true densities, porosity and humidity.

The most commonly used fillers of dry construction mixtures are limestone powder and fine quartz sand. With regards to technogenic materials the most commonly used filler is coal fly ash. Dry construction mixtures and mortars based on them containing these fillers are the most studied [74]. Mortars with regard to dry construction mixtures, containing stone powder from eruptive rocks - granite, basalt, and others, have been investigated to a lesser extent.

The purpose of our research was to determine the effectiveness of using aspiration granite dust (AGD) obtained by crushing granite into crushed stone as filler for dry construction mixtures and mortars based on them.

Granite belongs to magmatic intrusive rocks. The mineral composition of granites is dominated by alkaline feldspars, the content of which reaches 60 ... 65%. The second component of granite is quartz – 25 ... 35% [46]. Minerals composing granites, according to the classical approach, are chemically inactive in cement concrete with normal hardening and steaming. Just in autoclave processing they interact chemically with Portland cement. However, if in granites as well as in other acidic rocks, the part of silica is represented by an amorphous variety, they can be characterized by a certain puzzolan activity [18]. Amorphous silicon dioxide in the granite dust particles represents a nano-sphere sized from 200 to 600 nm. Active fillers include mineral powders with CaO absorption activity from 10 to 75 mg/g. According to available data, particles of granite siftings fraction < 0.16 mm had an activity of 18.3 mg/g of CaO and those of gabro-diabase - 16.52 mg/g of CaO.

The aspiration dust activity is largely characterized by the energy state of grain surface and crystalline lattice defects, formed during fragmentation due to free radicals as a result of the chemical bonds destruction. As is known [75], the surface of mineral materials is mosaic and is characterized by the sum of charges with a certain sign. For particles < 0.16 mm there is a significant surface amorphization level. They always contain adsorbed water and a layer of hydroxyl groups. The presence of these groups is due to occurrence of charges and jump of potential at the phases interface. The hydroxyl group distribution also depends on the acid-base (donor-acceptor) properties of the mineral powder surface. On the surface of silica-aluminate minerals, including granite dust, there is a significant amount of acidic centers [25], which should intensify the Portland cement hydration processes.

Research showed [3] that granite powder is highly hydrophilic, has a negative potential and, according to the existing classification, has a high-energy surface and, accordingly, an increased adsorption capacity. According to chemical theory [76], the adsorption capacity of solids is determined by the presence of unsaturated valences of atoms or molecules on their surface.

Increasing the activity of granite powder as mineral filler in cement composites is possible with an increase in its dispersion. For aspiration dust the dispersion depends on the method and mode of its capture as well as on dust collecting devices characteristics.

From the general theoretical considerations, in order to achieve high adhesion strength in the cement binding system, it is important to provide the required wettability of the filler by binder, which is possible by reducing the interphase surface energy when processing the filler by surface active substances. Dry construction mixtures include SAS which as a rule, are part of the additives group, regulating the rheological properties of mortar mixtures. In addition to plasticizers or superplasticizers, they include water-retaining (cellulose ethers, etc.), stabilizing admixtures (starch esters, etc.). If most plasticizers are part of ionic SAS group, then such additives as cellulose or starch esters are nonionic SAS.

In the dry construction mixtures (DCM) technology single-stage mixing of all components is usually used. For a more complete interaction of SAS with filler in order to increase its activity, two-stage mixing is more appropriate: at the first stage filler and SAS admixture, at the second - a modified filler, which can be considered as an organo-mineral modifier with other DCM components. Such a method can also be considered more promising from the viewpoint of achieving a high water-reducing effect of a superplasticizing SP admixture in mortar mixtures [77].

To increase the chemical activity of mineral powders, such as granite dust, which does not interact or interact slightly with $Ca(OH)_2$, it is promising to combine them with highly active silica or alumina silica additives.

7.2 Technological properties of mortar mixtures with AGD

To investigate the influence of the main factors of mortar mixtures composition with aspiration granite dust (AGD) and complex additives - polyfunctional modifiers (PFM), algorithmic experiments were carried out according to a typical three-level Ha_5 plan to obtain the statistical models of their workability by standard cone penetration (CP). Experiment planning conditions are given in Table 7.1. The models for mortar mixtures workability, obtained by statistical processing are presented in Table 7.2.

Analysis of the obtained models shows that superplasticizer content (factor X_2) and water demand (factor X_4) have the dominant effect on the modified mortar fluidity. There is a considerable effect of interaction between factors $X_{1\,(I)}$ and X_2, which show that the simultaneous change of these factors (the contents of individual polyfunctional modifier PFM_1 components) increases their integral action. Some interaction is between factors $X_{1\,(III)}$ and X_3, as well as between the factors X_3 and X_5.

Table 7.1 Experiments planning conditions for obtaining Eqs. 7.1–7.3.

Technological factors		Variation levels			Variation interval
Natural	Coded	−1	0	+1	
Air entraining admixture content, % of cement weight	$X_{1(I)}$	0	0.025	0.05	0.025
Water retaining admixture content (cellulose ether - EC), % of cement weight	$X_{1(II)}$	0	0.15	0.3	0.15
Content of metakaolin, % of cement weight	$X_{1(III)}$	0	7.5	15	7.5
Content of naphthalene-formaldehyde superplasticizer (SP), % of cement weight	X_2	0	0.35	0.7	0.35
F/C*	X_3	0	0.35	0.7	0.35
Water demand W, l/m^3	X_4	240	270	300	30
W/C	X_5	0.6	0.8	1.0	0.2

*Ratio between filler weight (AGD) to cement weight.

Table 7.2 Experimental-statistical models of modified mortar mixtures workability (cone penetration - CP).

Modifier	Regression equation	
PFM$_1$ (superplasticizer SP + air entraining admixture (AEA))	$CP_{(1)} = 9.73 + 1.08X_{1(I)} + 2.2X_2 + 0.62X_3 + 1.25X_4 -$ $1.14X_{1(I)}^2 + 0.37X_2^2 + 0.62X_3^2 + 0.37X_4^2 + 0.37X_5^2 +$ $0.28X_{1(I)}X_2 + 0.28X_2X_4 + 0.41X_3X_5 - 0.59X_4X_5$	(7.1)
PFM$_2$ (superplasticizer SP + water retaining admixture Tylose (EC))	$CP_{(2)} = 8.44 - 0.56X_{1(II)} + 2.47X_2 + 0.97X_4 + 0.07X_{1(II)}^2 +$ $0.82X_2^2 - 0.18X_3^2 - 0.18X_4^2 + 0.07X_5^2 - 0.25X_{1(II)}X_2 +$ $0.38X_2X_5 - 0.50X_3X_4 - 0.69X_4X_5$	(7.2)
PFM$_3$ (superplasticizer SP + metakaolin (MK))	$CP_{(3)} = 9.06 - 0.28X_{1(III)} + 2.36X_2 + 0.53X_3 + 1.03X_4 -$ $0.51X_{1(III)}^2 + 0.24X_2^2 - 0.24X_3^2 + 0.24X_4^2 + 0.24X_5^2 +$ $0.38x_{1(III)}X_3 + 0.44X_3X_5 - 0.44X_3X_4 - 0.89X_4X_5$	(7.3)

All mixtures had a constant cement to sand ratio by weight (1:3).

In the obtained polynomial models of cone penetration in the mortar mixture with PFM admixtures is an insignificant influence of factor X_5 (water-cement ratio) - in the range 0.6 ... 1.0. From concrete technology it is known that to a certain critical W/C or C/W the water constancy rule is preserved, that is, with the change of these parameters to a certain critical value, the water demand remains practically constant [65]. Obviously, in this case the water demand constancy rule at different contents of PFM and F/C components in mortar mixtures is also valid. At the same time, along with the filled paste volume, the ratio of aspiration dust to cement (F/C) and the metakaolin content significantly affect the fluidity.

Other experiments were carried out to study the change in the workability of mixtures with PFM additives in time. Experiments were carried out at a temperature of 20 ± 2°C. The mixtures had W/C = 0.6, the type and content of individual PFM components were varied. Initial fluidity by standard cone penetration was 11 ... 13 cm. The experimental results are shown in Table 7.3.

The results demonstrate that the mixture consistency (cone penetration depth > 10 cm) in mixtures with the addition of SP (0.3%) without AGD is kept for 20 minutes, with the addition of SP (0.3%) + AED (0.03%) - 60 min., with the addition of SP (0.3%) + EC (0,3%) - 100 min. Adding AGD at F/C < 0.4 positively affects the workability preservation, with a further increase in F/C, the workability decreases. The positive influence of AGD on the stabilization of mortar mixtures workability in presence of SP can be associated with their increased water-retaining capacity.

Air entraining and especially water retaining admixtures have a stabilizing effect on fluidity loss. They yield a significant increasing stability in the time of mortar mixtures containing superplasticizer.

Table 7.3 Changing the workability of mortar mixtures with different filler:cement ratios and PFM admixtures in time.

Composition No.	F/C	Cone penetration, cm, after, min.				
		20	40	60	80	100
SP (0.3%)						
1	0	10.8	7.3	5.2	2.4	-
2	0.35	11.4	8.5	6.1	3.7	-
3	0.7	9.5	6.5	3.7	1.5	-
PFM$_1$: SP (0.3%) + AEA 0.03%						
4	0	12.2	10.5	9.9	6.8	3.2
5	0.35	11.6	10.6	10.2	7.5	4.5
6	0.7	11.2	10.0	8.6	5.8	3.4
PFM$_2$: SP (0.3%) + EC 0.3%						
7	0	11.6	10.7	10.4	10.1	6.9
8	0.35	11.5	10.6	10.5	10.3	8.1
9	0.7	10.5	10.4	10.0	8.7	6.4
PFM$_3$: SP (0.3%) + MK 10%						
10	0	10.5	6.8	3.4	1.8	-
11	0.35	10.9	7.5	5.6	2.9	-
12	0.7	9.8	5.8	4.9	1.8	-

Mixtures containing metakaolin lose their fluidity faster than without it. Kinetics of the fluidity loss of modified mortars agrees with their setting time and the plastic strength of the filled cement paste.

The curves of fluidity for mortar mixtures with PFM admixtures can be divided into stages relative to stability and progressive decrease. Experimental data show that AGD admixture significantly affects the rate of fluidity loss.

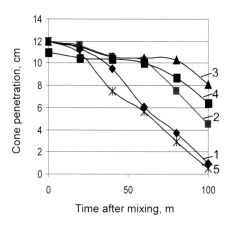

Fig. 7.1 Dependences of mortar mixtures fluidity on time after mixing (F/C = 0.35):
1 – SP=0.3%; 2 – SP (0.3%) + AEA 0.03%; 3 – SP (0.3%) + EC 0.3%;
4 – SP-1 (0.3%) + EC 0.3%; 5 – SP-1 (0.3%) + metakaolin 10%.

The kinetics of the mortar mixtures rheological properties change in time essentially depends on the temperature factor. Increasing the temperature to 30°C practically does not lead to a drop in the initial mortar mixtures fluidity. At temperatures above 30°C, the period of relative stability for fluidity of mixtures with all investigated PFM (excepted PFM$_2$) practically disappears, although the initial fluidity with PFM admixture is higher. Thus, at 40° there occurs a decrease in the initial fluidity of mortar mixtures up to 6 ... 7 cm without admixtures and up to 7 ... 8 cm for mixtures with PFM. At 60°C the stability in time of fluidity decreases to 3 ... 4 cm, which practically makes the use of mortars impossible.

Analysis of experimental data shows that the mortar mixtures fluidity in time in general terms varies according to the equation:

$$CP_\tau = CP_0\left(1 - k\tau\right) \tag{7.4}$$

where CP_0 is fluidity (cone penetration) of mortar mixture, obtained immediately after mixing ($\tau = 0$); CP_τ is fluidity of mortar mixture at τ hours after mixing; k is coefficient of relative change in the mixture fluidity, which depends on the temperature of the ambient air and the type of PFM.

The changes in the fluidity in time, obviously depends on all the factors that determine the rate of cement hydration. In addition to W/C and temperature, such factors include cement activity and dispersion, type and content of admixtures.

7.3 Adhesion properties of mortars with AGD

One of the main tasks of modifying construction mortars is to increase their adhesion ability, which contributes to the improving of their quality and durability. The most important task is providing adhesive properties for gluing and masonry mortars.

At present, there is no generally accepted theory that would explain the gluing process satisfactorily. Due to specificity and variety of phenomena that occur at different gluing process stages creating of general bonding theory is considerably complicated [78].

Along with the adding polymeric admixtures, there are many ways to improve the adhesive ability of cement stone at its limited content in concrete. One of them is based on a concept that considers cement stone as micro-concrete [79]. Following this concept it is expedient to increase the cement glue dispersion, ensuring its full hydration. Cement grain > 40 microns, which are practically not hydrated, should be replaced by fillers. This concept forms a basis for technologies of dry and additional wet milling of cement with sand and other fillers in order to obtain colloidal cement glue [78]. Additional milling of cement, however, is not widely used due to high energy consumption, imperfect design of milling aggregates, rapid loss of binder activity. Using colloidal cement glue, obtained by vibrating mills and mixers, this is limited to a very narrow area of glue mixtures.

Adhesive ability of cement mortars was determined as peel strength from the concrete base of a 50 × 50 mm ceramic tile specimen. The influence of water-cement ratio (W/C) and the filler - cement ratio (F/C), as well as the type and content of the investigated admixtures, were studied using experiments according to B_4 plan (Table 7.4).

Mathematical processing of the results enabled to obtain experimental-statistical models for adhesion strength (7.5 ... 7.10) that are given in Table 7.5. To analyze the

Table 7.4 Experiment planning conditions for obtaining Eqs. 7.5 ... 7.16.

Technological factors		Variation levels			Variation interval
Natural	Coded	−1	0	+1	
W/C	X_1	0.6	0.8	1.0	0.2
F/C	X_2	0	0.35	0.7	0.35
Content of superplasticizer SP , % of cement weight	X_3	0	0.35	0.7	0.35
Content of air entraining admixture, % of cement weight	$X_{4(I)}$	0	0.025	0.05	0.025
Content of water retaining admixture Tylose (EC), % of cement weight	$X_{4(II)}$	0	0.15	0.3	0.15
Metakaolin content	$X_{4(III)}$	0	5	10	5

Table 7.5 Mathematical models of cement mortars adhesion strength.

Admixture	Mathematical model of adhesive strength	
PFM$_1$ (SP +AEA)	$f_{ao}^7 = 0.356 - 0.028x_1 + 0.035x_3 + 0.017x_{4(1)} - 0.0459x_1^2 - 0.0459x_2^2 - 0.0209x_3^2 - 0.0459x_{4(1)}^2$	(7.5)
	$f_{ao}^{28} = 0.639 - 0.028x_1 + 0.018x_2 + 0.057x_3 + 0.026x_4 - 0.0792x_1^2 - 0.074x_2^2 - 0.0542x_3^2 - 0.0542x_{4(1)}^2 + 0.011x_2x_3$	(7.6)
PFM$_2$ (SP + EC)	$f_{ao}^7 = 0.392 - 0.027x_1 + 0.036x_3 + 0.034x_{4(II)} - 0.0574x_1^2 - 0.0424x_2^2 - 0.0224x_3^2 - 0.0424x_{4(II)}^2$	(7.7)
	$f_{ao}^{28} = 0.706 - 0.03x_1 + 0.021x_2 + 0.06x_3 + 0.052x_{4(II)} - 0.097x_1^2 - 0.067x_2^2 - 0.067x_3^2 - 0.067x_{4(II)}^2 + 0.016x_2x_3$	(7.8)
PFM$_3$ (SP+ MK)	$f_{ao}^7 = 0.368 - 0.029x_1 + 0.0336x_3 + 0.02x_{4(III)} - 0.0382x_1^2 - 0.0482x_2^2 - 0.0132x_3^2 - 0.0132x_{4(III)}^2$	(7.9)
	$f_{ao}^{28} = 0.666 - 0.034x_1 + 0.021x_2 + 0.058x_3 + 0.035x_{4(III)} - 0.0767x_1^2 - 0.082x_2^2 - 0.0517x_3^2 - 0.0317x_{4(III)}^2 + 0.013x_2x_3$	(7.10)

obtained models, two-factor graphic dependences and response surface of adhesive strength were constructed (Figs. 7.2 ... 7.5).

Additionally, the effect of increase in AGD specific surface and the ratio between the filled cement binder and filler on the adhesive strength magnitude were studied. The results of experiments are shown in Table 7.6 and Fig. 7.7.

Analysis of the obtained adhesion strength models shows that the impact on it of both W/C and F/C has an extreme nature (Figs. 7.2 ... 7.3). Thus, with an increase in mortar W/C from 0.6 to 0.7 ... 0.75, the adhesive strength increases by 8 ... 10%, with a further increase in W/C, it decreases by 20 ... 25% of the maximum value. At the same time the maximum of f_{ad} is observed at F/C = 0.35 ... 0.4. Such an effect of the mentioned technological factors on adhesion can be explained by the change of the contact layer porosity and also the wetting degree of the base by mortar.

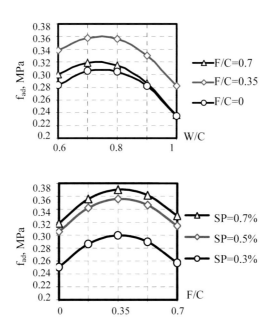

Fig. 7.2 Dependence of adhesive strength (MPa) of cement mortars, modified by PFM$_1$ on W/C and F/C at 7 days.

Superplasticizer SP and air-entraining admixture as a SAS, have a positive effect on the mortars adhesion (Figs. 7.2, 7.3) as a result of changes in their surface energy and the contact layer qualitative characteristics. Addition of a hydrophilic SP admixture improves the mortar contact layer characteristics, obviously, primarily, increasing its wettability and reducing the moisture content excess. With an increase in the superplasticizer content from 0 to 0.35% of the cement weight, adhesion strength increases by 16 ... 35% at W/C = 0.6 and 25 ... 40% at W/C = 1.0. Further increase in SP content to 0.7% yields an increase in f_{ad} by 13 ... 20%.

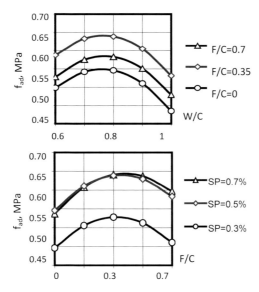

Fig. 7.3 Dependence of adhesive strength (MPa) of cement mortars, modified by PFM$_1$, on W/C and F/C at 28 days.

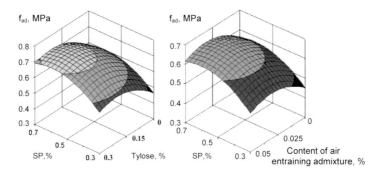

Fig. 7.4 Dependence of adhesive strength (MPa) of cement mortars at 28 days on modifying admixtures content (F/C = 0.35, W/C = 0.8).

Addition of air entraining and especially the polymeric admixture Tylose provides the necessary water retaining ability of the mortar mixture and reduces the required adhesive layer thickness. At other equal conditions, increase in the content of the air entraining admixture from 0 to 0.025% by cement weight adhesive strength increases by 30 ... 45%, with further increase in the AEA content the adhesive strength is slightly reduced. Increase in Tylose content from 0 to 0.15% leads to an increase in adhesion strength by 25 ... 55%, further increase in the content of the admixture weakly affects the mortars adhesion.

Adding metakaolin increases the volume of hydrated compounds which also positively affects the adhesion strength of mortars (Fig. 7.5). An increase in the MK content up to 10% of the cement weight leads to an increase in adhesion strength by 16 ... 33% in all terms of hardening at other equal conditions.

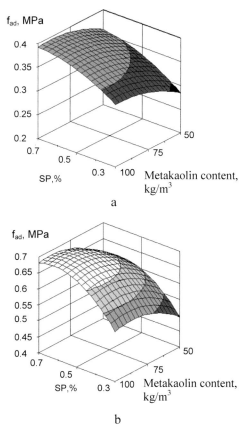

Fig. 7.5 Adhesive strength (MPa) of cement mortars with metakaolin admixture (MK) at: a – 7 days; b – 28 days.

An additional factor contributing to increase in mortars adhesion strength is an increase in the AGD dispersion (Table 7.6, Fig. 7.6). Thus, an increase in its specific surface (S_s) from 290 to 390 m²/kg can increase the bond strength of the mortar with the base at 7 days by 14 ... 26% and at 28 days - by 13 ... 18%.

It is known that new freshly formed surfaces of mineral materials have considerably higher surface energy values, which determines their higher adhesion activity. Increase in the surface area and, correspondingly, in the surface energy, causes an increase in the isobaric potential of the powders and, accordingly, their chemical activity, which also contributes to high adhesion strength at their contact with binder [80]. Adsorption of high dispersion powders of water vapor and carbon dioxide from the air and saturation of unbalanced molecular forces leads to the fillers surface "aging" and also becomes an additional barrier for formation of reliable adhesive contact.

The change in the cement binder – sand ratio also affects the adhesion strength (Fig. 7.6). Increasing the binder – aggregate ratio leads to an increase in adhesion strength which is more pronounced for sand with a higher grain size.

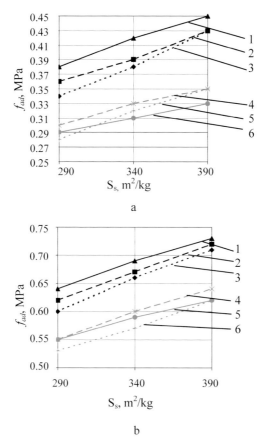

Fig. 7.6 Dependence of modified cement mortars adhesive strength on AGD dispersion and binder: sand ratio: a – at 7 days, b – at 28 days: 1 ... 3 – (C+AGD) : Sand = 1 : 3; 4...6 – (C+AGD) : Sand = 1 : 4.5; 1, 6 – EC = 0.15%; 2, 4 – MK = 7.5%; 3, 5 –AEA = 0.025%.

7.4 Strength and deformation properties of modified mortars using AGD

The operational reliability of the masonry and adhesive mortars, along with their adhesion to the surfaces to which they are applied, is characterized by the cohesion, which determines the mortar strength. Investigation of the modified mortars strength was carried out by implementing algorithmic experiments with planning conditions shown in Table 7.4. Beam specimens, which hardened under normal conditions, were tested at 28 days. Statistical processing of the experimental results yielded regression equations for compressive strength (f_c) and flexural strength ($f_{c.tf}$), given in Table 7.7, and the graphic dependencies that illustrate them in Figs. 7.7 ... 7.10.

Analysis of the obtained equations and graphs shows that for the investigated cement mortars W/C has a dominant influence on both compression and flexural strength. The next important factor is the superplasticizer content. The impact of F/C

Table 7.6 Adhesive strength of modified cement mortars using AGD with different dispersion.

(C+AGD)/Sand	S_{AGD}	Adhesive strength, MPa, at	
		7 days	28 days
Modifier PFM$_1$ (W/C = 0.8; F/C = 0.6; SP = 0.5%; AEA= 0.03%)			
1 : 3	290	0.34	0.60
	340	0.38	0.66
	390	0.43	0.71
1 : 4.5	290	0.28	0.53
	340	0.32	0.57
	390	0.35	0.62
Modifier PFM$_2$ (W/C = 0.8; F/C = 0.6; SP = 0.5%; MK = 0.3%)			
1 : 3	290	0.38	0.64
	340	0.42	0.69
	390	0.45	0.73
1 : 4.5	290	0.29	0.55
	340	0.31	0.59
	390	0.33	0.62
Modifier PFM$_3$ (W/C = 0.8; F/C = 0.6; SP= 0.5%; MK = 5%)			
1 : 3	290	0.36	0.62
	340	0.39	0.67
	390	0.43	0.72
1 : 4.5	290	0.30	0.55
	340	0.33	0.60
	390	0.35	0.64

on both compressive and flexural strength is extreme. For all investigated compositions close to optimal value of F/C is 0.35 ... 0.4. The highest influence of F/C is observed for mortars with complex filler containing aspiration dust and metakaolin.

Positive influence of fillers on strength of concrete and mortars is characterized by "cementing efficiency" [18], determined as cement saving, achieved by adding into mortar 1 kg of filler at constant W/C. For filled systems compressive strength can be calculated using the following expression [13], which is a modified well-known formula

$$f_c = AR_c \left(\frac{C + C_{c.e}F}{W} \right) \qquad (7.17)$$

where $C_{c.e}$ is the cementing efficiency coefficient; F is the weight of filler, kg/m³, R_c – cement compressive strength, MPa.

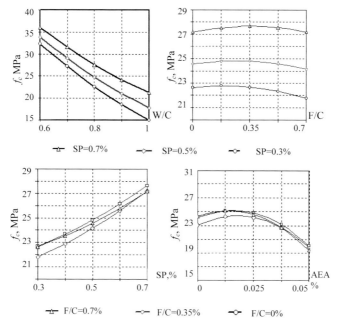

Fig. 7.7 Dependences compression strength of cement mortars, modified by PFM₁ admixture.

Table 7.7 Mathematical models of modified cement mortars strength.

Admixture	Regression equation	
PFM₁ (SP-+AEA)	$f_c = 24.8 - 8.04x_1 + 1.207x_2 + 2.49x_3 - 2.03x_4 + 1.04x_1^2 - 0.458x_2^2 + 0.342x_3^2 + 2.71x_4^2 + 0.33x_1x_2 + 0.66x_1x_3 + 0.23x_2x_3 + 0.44x_2x_4 - 0.18x_3x_4$	(7.11)
	$f_{c.tf} = 3.74 - 0.986x_1 + 0.353x_3 - 0.252x_4 + 0.401x_1^2 - 0.151x_2^2 - 0.051x_3^2 - 0.449x_4^2 - 0.044x_1x_2 + 0.056x_1x_4$	(7.12)
PFM₂ (SP-+EC)	$f_c = 25.5 - 8.4x_1 + 1.106x_2 + 2.61x_3 - 1.62x_4 + 1.83x_1^2 - 1.62x_2^2 + 1.03x_3^2 - 1.87x_4^2 + 0.32x_1x_2 + 1.77x_1x_3 + 1.21x_2x_3 - 0.18x_3x_4$	(7.13)
	$f_{c.tf} = 3.93 - 0.935x_1 - 0.017x_2 + 0.291x_3 - 0.289x_4 + 0.459x_1^2 - 0.267x_2^2 - 0.491x_4^2 - 0.069x_1x_2 - 0.044x_1x_4 - 0.031x_3x_4$	(7.14)
PFM₃ (SP+MK)	$f_c = 26.5 - 7.84x_1 + 1.89x_2 + 2.44x_3 + 0.986x_4 + 0.948x_1^2 - 0.452x_2^2 + 0.998x_3^2 - 0.652x_4^2 + 0.294x_1x_3 + 0.506x_1x_4 + 0.858x_2x_3 + 1.14x_2x_4 + 0.41x_3x_4$	(7.15)
	$f_{c.tf} = 3.92 - 0.862x_1 + 0.224x_2 + 0.336x_3 + 0.146x_4 + 0.273x_1^2 + 0.173x_2^2 + 0.173x_3^2 - 0.2272x_4^2 + 0.063x_1x_2 - 0.038x_1x_3 - 0.1x_2x_3 + 0.038x_2x_4$	(7.16)

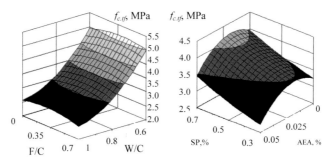

Fig. 7.8 Dependences of flexural strength of cement mortars, modified by PFM₁ admixture.

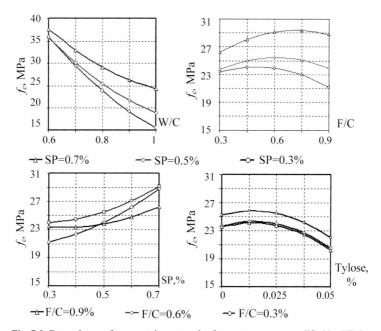

Fig. 7.9 Dependence of compressive strength of cement mortars, modified by PFM₂.

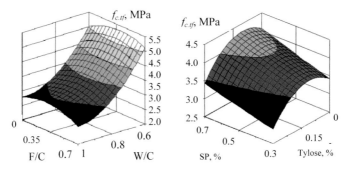

Fig. 7.10 Dependences of flexural strength of cement mortars, modified by PFM₂.

As follows from the obtained data, the value of $C_{c.e.}$, essentially depends on the F/C ratio, and at a given cement consumption - on the filler content. According to Figs. 7.7 and 7.10, the positive effect of AGD on the mortars strength can be significantly reduced at non-optimal F/C ratio and AGD content.

Influence of polyfunctional modifiers (PFM) on mortars strength depends on their composition and content. At constant mortar mixtures fluidity, the increase in content of superplasticizer in the PFM composition naturally leads to strength growth. In particular, due to increase in SP content from 0.3 to 0.7%, the mortar compressive strength becomes higher by 23 ... 34% and flexural strength - by 18 ... 24% at other equal conditions. Increase in the content of water-retaining and, especially, air-entraining admixture leads to a decrease in strength by 10 ... 14% and 14 ... 21% respectively. The most noticeable decrease in strength is observed at the content of water-retaining admixture Tylose more than 0.3%, and air-entraining - more than 0.04% of the cement weight. However, when assessing the impact of these additives, it is worth it to consider that when casting on a porous basis, negative influence of these additives on modified mortars strength is offset by increased workability of mortar mixtures and high mortar seam quality. Addition of metakaolin, taking active part in the hydration and structuring processes, positively affects the strength of mortars containing AGD, at adding superplasticizer to the mixture.

Analysis of models (7.16 and 7.17) shows that the increase in MK content in the binder composition to a certain limit leads to a marked increase in mortar strength at close cement-water ratio values. With an increase in the MK content from 30 to 50 kg/m^3, the mortar compressive strength increases by 13 ... 20%, depending on the AGD and superplasticizer content, and the flexural - by 9 ... 15%. An increase in MK content over 50 kg/m^3 is not feasible, since further increase in strength is not observed, in addition, with high MK content water demand of mortars increases significantly.

Modification mortars by PFM positively affects the ratio between tensile and compression strength, which is one of the criteria for mortars cracking resistance. Figures 7.11 and 7.12 show data that illustrates the change in $f_{c.tf}/f_c$ up to 180 days for mortars modified by PFM.

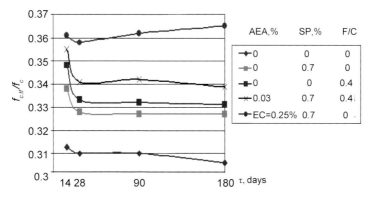

Fig. 7.11 Dependence of cracking resistance criterion ($f_{c.tf}/f_c$) on hardening duration of mortar, modified by PFM$_1$ and PFM$_2$ (W/C = 0.7).

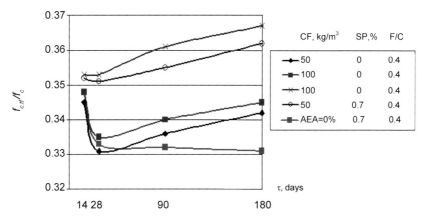

Fig. 7.12 Dependence of cracking resistance criterion $(f_{c.tf}/f_c)$ on hardening duration of mortar, modified by PFM$_3$ (W/C = 0.7) (CF – carbonate filler).

As it follows from the experimental data, modification of mortars by organic and organo-mineral additives positively affects the crack resistance criterion. Increase in its values is characteristic for modified mortars at 28 days, as well as in other hardening terms.

The deformation properties of mortars and concrete are determined by the modulus of elasticity. The modulus of elasticity E of mortar or concrete can be predicted based on the theoretical dependence, obtained when considering it as a two-phase system with spherical particles of aggregate, uniformly distributed in the cement stone [81,82]:

$$E = \frac{V_{st} + (2 - V_{st}) E_a / E_{st}}{2 - V_{st} + V_{st} E_a / E_{st}} \qquad (7.18)$$

where V_{st} is the volumetric concentration of cement stone; E_a and E_{st} are modulus of elasticity of aggregate and cement stone, respectively.

Porosity V_p and elasticity of the so-called "gel-crystalline phase" have a dominant influence on the cement stone modulus of elasticity, which follows from the famous expression [18]:

$$E_{st} = (1 - V_p)^3 E_g \qquad (7.19)$$

Available data on the effect of SAS are controversial, although most researchers [18] believe additives reduce the modulus of elasticity of the concrete. The decrease in modulus of elasticity of cement composites with SAS is due to the adsorption modification of the cement stone structure or appearance in hydrated shells in grains of more structural elements, which leads to an increase in the particles slip surface. When adding plasticizers it is necessary to take into account, along with the structure modification, the effect of two possible opposite effects - a decrease in the cement stone porosity and a decrease of the cement stone volume in concrete. When using a superplasticizer to reduce W/C and increase strength, the concrete or mortar modulus of elasticity can significantly increase.

Polymer additives can also lead to some increases in the modulus of elasticity. Following I. Ohama [83], at polymer cement ratio of 0.05, addition of polyvinyl acetate dispersion increased the concrete modulus of elasticity from 2.11×10^4 MPa to 2.27×10^4 MPa.

The modulus of elasticity of masonry mortars with additives-modifiers was obtained experimentally at 28 days for $70 \times 70 \times 280$ mm specimens. Deformations were measured using strain gauges with an accuracy of 0.7×10^{-5}. The prisms were loaded in spring units with steps of $0.05\, f_{c.pr}$ with a hold of 5 minutes up to $0.3\, f_{c.pr}$ ($f_{c.pr}$ is the prismatic strength of the mortar). The modulus of elasticity values were calculated by the equation:

$$E = \sigma / \varepsilon \qquad (7.20)$$

where σ is the applied stress; ε is the longitudinal deformation.

The values of strength and modulus of elasticity of the investigated masonry mortars are given in Table 7.8, strength and modulus of elasticity of glue mortars are given in Table 7.9. The cement - sand ratio is taken from the condition of achieving a mortar fluidity, which corresponds to cone penetration of 7 ... 9 cm.

The analysis of data in Table 7.8 shows that the prismatic strength of the investigated mortars is within the range of $(0.81 ... 0.88)\, f_c$ and it can be roughly found from equation

$$f_{c.pr} = 0{,}85\, f_c, \qquad (7.21)$$

where f_c is the mortar compressive strength.

As known, for normal-weight concrete there is a correlation dependence $f_{c.pr} = 0.783 f_c$ [82]. The above mentioned equations of prismatic and cubic strength are obtained for mortars and sand concrete by the other researchers too [33].

When predicting the concrete modulus of elasticity loaded at age τ days the most often used dependencies have the following type [4.11]:

$$E = \frac{E_m f_c}{K + f_c} \qquad (7.22)$$

where f_c is the compressive strength at age τ; E_m and K are empirical constants. Most building codes recommend taking $E_m = 52000$ and $K = 23$.

Analysis of experimental values of mortar's modulus of elasticity of mortars (Table 7.8) shows that they are approximately 20% lower than the calculated according to Eq. (7.22) for concrete.

Table 7.8 Influence of polyfunctional modifiers on prismatic strength and modulus of elasticity of mortars.

No.	W/C	F/C	Content of AEA or EC, % of binder weight	Content of SP, % of binder weight	Compressive strength, MPa, at 28 days		Modulus of elasticity $E \times 10^4$ MPa
					Cubic	Prismatic	
Cement mortar with PFM$_1$ (SP+AEA)							
1	0.6	-	-	-	26.9	22.7	2.11
2	0.6	-	-	0.7	28.7	24.2	2.26
3	0.6	0.4	-	-	31.3	27.2	2.45
4	0.6	0.4	0.03	0.7	29.8	25.0	2.33
5	0.7	-	-	-	24.9	20.1	2.11
6	0.7	-	-	0.7	26.2	22.2	2.16
7	0.7	0.4	-	-	27.8	24.7	2.28
8	0.7	0.4	0.03	0.7	25.7	22.2	2.04
9	0.8	-	-	-	23.0	19.4	2.00
10	0.8	-	-	0.7	25.6	21.7	2.07
11	0.8	0.4	-	-	26.4	22.6	2.21
12	0.8	0.4	0.03	0.7	24.9	21.2	2.03
Cement mortar with PFM$_2$ (SP + EC)							
13	0.6	0.4	0.25	0.7	32.9	28.0	7.3
14	0.7	0.4	0.25	0.7	30.6	26.2	8.0
15	0.8	0.4	0.25	0.7	28.5	24.0	8.5
Cement mortar with AGD and metakoalin (MK = 10%)							
16	0.6	0.4	-	-	26.9	22.7	2.11
17	0.6	0.4	-	0.7	28.7	24.2	2.26
18	0.7	0.4	-	-	31.3	26.9	2.45
19	0.7	0.4	-	0.7	29.8	24.7	2.33
20	0.8	0.4	-	-	24.9	20.2	2.11
21	0.8	0.4	-	0.7	26.2	22.0	2.16

Table 7.9 Dynamic modulus of elasticity and conditional deformability (CD) of glue mortars.

No.	Mortar composition				$E_d \times 10^{-4}$, MPa, at			CD $\times 10^{-4}$, at		
	W/C	F/C	W, kg/m^3	MK/F	28 days	90 days	180 days	28 days	90 days	180 days
1	0.8	-	250	-	3.63	3.98	4.17	1.13	1.12	1.10
2	0.75	0.4	270	-	3.25	3.77	3.96	1.08	1.10	1.10
3	0.82	0.4	277	0.1	3.83	4.44	4.58	1.22	1.28	1.30
4	0.85	0.4	281	0.3	3.90	4.45	4.56	1.24	1.30	1.33

A comprehensive assessment of concrete deformability and crack resistance is the ratio of tensile or flexural strength to the static or dynamic modulus of elasticity. These parameters are rather close.

As known [13], for fine-grained concrete, the equation for calculating splitting tensile strength (f_{cts}) and flexural tensile strength (f_{ctf}) differ just by coefficients

$$f_{cts} = 0.57\, f_c^{2/3}, \quad f_{c.tf} = 0.99\, f_c^{2/3}. \tag{7.23}$$

Comparison of experimental values of static and dynamic modules of elasticity shows that the ratio E/E_d for concrete and mortars is in the range of 0.87–0.95. Lower values are characteristic for a material with compression strength of less than 25 MPa

Table 7.10 Conditional deformability of modified masonry mortars.

No.	W/C	F/C	Content of AEA or EC, % of binder weight	Content of SP% of binder weight	Flexural strength f_{ctf}, MPa	$\dfrac{f_{cf}}{f_c}$	Modulus of elasticity $E \times 10^4$ MPa	Conditional deformability, $CD \times 10^{-4}$
\multicolumn{9} Cement mortar with PFM$_1$ (SP+AEA)
1	0.6	-	-	-	8.34	0.31	2.11	2.06
2	0.6	-	-	0.7	9.76	0.34	2.26	2.25
3	0.6	0.4	-	-	10.02	0.32	2.45	2.13
4	0.6	0.4	0.03	0.7	9.83	0.33	2.33	2.19
5	0.7	-	-	-	7.72	0.31	2.11	1.90
6	0.7	-	-	0.7	8.65	0.33	2.16	2.08
7	0.7	0.4	-	-	9.17	0.33	2.28	2.09
8	0.7	0.4	0.03	0.7	8.74	0.34	2.04	2.23
9	0.8	-	-	-	7.36	0.32	2.00	1.91
10	0.8	-	-	0.7	8.70	0.34	2.07	2.19
11	0.8	0.4	-	-	8.71	0.33	2.21	2.05
12	0.8	0.4	0.03	0.7	8.96	0.36	2.03	2.30
\multicolumn{9} Cement mortar with PFM$_2$ (SP + EC)
13	0.6	0.4	0.25	0.7	11.84	0.36	2.57	2.40
14	0.7	0.4	0.25	0.7	11.02	0.36	2.44	2.35
15	0.8	0.4	0.25	0.7	9.98	0.35	2.34	2.22
\multicolumn{9} Cement mortar with PFM$_3$ (AGD and metakoalin (MK = 10%))
16	0.6	0.4	-	-	8.88	0.33	2.11	2.19
17	0.6	0.4	-	0.7	10.05	0.35	2.26	2.31
18	0.7	0.4	-	-	10.02	0.33	2.45	2.13
19	0.7	0.4	-	0.7	10.43	0.35	2.33	2.33
20	0.8	0.4	-	-	8.22	0.33	2.11	2.03
21	0.8	0.4	-	0.7	8.91	0.34	2.16	2.14

[13]. The ratio f_{cts}/E_d is close to the limit elongation value and is called the conditional deformability *CD*. It's easy to show that

$$CD = \frac{f_{cts}}{E_d} \approx \frac{0.57 f_{c.tf}}{0.99 \cdot 1.1E} \approx 0.52 \frac{f_{c.tf}}{E} \qquad (7.24)$$

Tables 7.9 and 7.10 present the values of CD for the investigated mortars at 28 days, obtained according to Eq. (7.24).

The conditional elongation of unmodified mortars with an increase in hardening duration tends to decrease, while for modified ones it increases. Obviously, the sign of change in the conditional mortar deformability and, consequently, crack resistance with age is associated with the nature of the change in the tensile strength ($f_{c.t}$) and compressive strength (f_c). For mortars with PFM additives this ratio is higher than for unmodified ones at both 28 days, and at later hardening terms. A higher ratio of tensile to compressive strength for concrete and mortar with active mineral admixtures is noted in the works of other researchers [84].

To measure shrinkage, benchmarks in the end face of the prism were set 8 hours after compacting them in molds. The specimens were stored at $20 \pm 2°C$ and relative humidity of $75 \pm 5\%$. Shrinkage deformations curves are shown in Figs. 7.13 and 7.14.

Their analysis shows that for all investigated compositions shrinkage deformations are stabilized before 100 days. The highest shrinkage values have mortars without PFM additives. Their shrinkage up to the attenuation time ranges from 0.7 to

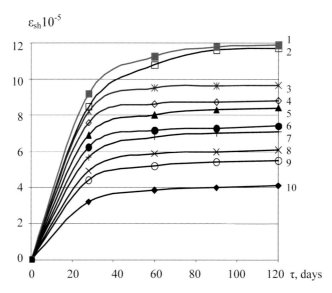

Fig. 7.13 Shrinkage deformations of modified cement mortars 1-5, 8 – W/C = 0.8: 1 – mortar without PFM, F/C = 0.4; 2 – mortar without PFM, F/C = 0; 3 – F/C = 0.4, SP = 0.7%, EC = 0.1%; 4 – F/C = 0.4, SP = 0.7%, EC = 0.4%; 5 – F/C = 0.4, SP = 0.7%, AEA = 0.01%; 6, 7, 9, 10 – W/C = 0.6: 6 – mortar without PFM, F/C = 0.4; 7 – mortar without PFM, F/C = 0; 8 – F/C = 0.4, SP = 0.7%, AEA = 0.05%; 9 – F/C = 0.4, SP = 0.7%, EC = 0.4%; 10 – F/C = 0.4, SP-1 = 0.7%, AEA = 0.04%

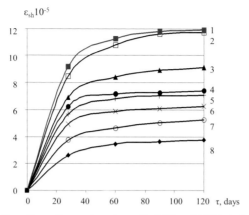

Fig. 7.14 Shrinkage deformations of cement mortars with AGD and MK 1-3, 6 – W/C=0.8: 1 – mortar without PFM, F/C=0.4; 2 – mortar without PFM, F/C=0; 3 – F/C=0.4, SP-1=0.7%, MK=50 kg/m³; 6 – F/C=0.4, SP=07%, MK=70 kg/m³; 4, 5, 7, 8 – W/C=0.6; 4 – mortar without PFM, F/C=0.4; 5 – mortar without PFM, F/C=0.

1.5 mm/m, which is characteristic for fine-grained concrete with similar W/C [33]. At early hardening stages the AGD additive increases mortars' shrinkage deformations, which later align with those in mortars without filler. PFM_2 admixture reduces the marginal shrinkage deformations by 20 ... 35% whereas PFM_1 - by 30 ... 50%. The degree of shrinkage deformation reduction increases as the PFM plasticization effect becomes higher. The most significant decrease in shrinkage is observed when adding AEA - by 40 ... 60%.

7.5 Frost resistance of mortars with AGD

Frost resistance of mortar and concrete is, like strength, determined by the nature of its porous structure. Capillary macropores are the main defect in the structure of compacted concrete and mortars, which reduces its frost resistance. According to available data, concrete is frost-resistant if the contents of capillary pores is up to 5 ... 7% [61]. A known empirical equation interrelating the frost resistance F and the capillary porosity of P_{cap} is:

$$F = \left(14 - P_{cap}\right)^{2.7} \tag{7.25}$$

The frost resistance significantly depends on pore sizes. Pores greater than 10^{-5} cm are filled by water and reduce the frost resistance of porous material.

In addition to the pore size, the most important parameter that determines the frost resistance is the ratio between the volumes of conditionally closed pores formed as a result of contraction and air entraining and open pores that are saturated with freezing water. This ratio can serve as a criterion of frost resistance [85].

To investigate the frost resistance of modified mortars, algorithmic experiments were performed in accordance with B_4 plan. Experiment planning conditions are given in Table 7.11.

The kinetics of changes in mortar strength through cyclic freezing and thawing were determined by testing the control and basic specimens and obtaining the frost resistance coefficient. Tests were performed every 25 cycles of freezing and thawing. By statistical processing of the obtained experimental data, polynomial models of modified mortars frost resistance were obtained (Table 7.12). Corresponding graphical analysis is presented in Figs. 7.15 ... 7.18.

Analysis of the data shows that the investigated mortars grade by frost resistance ($K_f \geq 0.95$) varies between F25 and F250. The highest frost resistance has mortar with air entraining admixture. The rate of strength loss in such mortars with freezing and thawing was also lower than for concrete without admixture. Positive effect on mortars frost resistance, though to a lower extent, have adding SP and cellulose ether. For filled

Table 7.11 Experiment planning conditions for obtaining Eqs. 7.26 and 7.27.

Technological factors		Variation levels			Variation interval
Natural	**Coded**	**−1**	**0**	**+1**	
W/C	X_1	0.6	0.8	1.0	0.2
F/C	X_2	0	0.35	0.7	0.35
Content od superplasticizer SP, % of binder weight	X_3	0	0.35	0.7	0.35
Content of air entraining admixture (AEA), % of binder weight	$X_{4(I)}$	0	0.025	0.05	0.025
Content of water retaining admixture (EC), %	$X_{4(II)}$	0	0.15	0.3	0.15
Content of metakaolin, % of cement weight	$X_{4(III)}$	0	7.5	15	7.5

Table 7.12 Mathematical models of modified mortars' frost resistance.

PFM type	Mathematical models of frost resistance	
SP + AEA	$F = 149.9 - 67.1x_1 - 4.46x_2 + 11.6x_3 + 28.7x_{4(I)}$ $- 36.6x_1^2 - 2.95x_2^2 + 1.95x_3^2 + 3.95x_{4(I)}^2 + 1.125x_1x_2 -$ $11.6x_1x_{4(I)} - 2.37x_2x_{4(I)} + 2.75x_3x_{4(I)}$	(7.26)
SP+EC	$F = 135.8 - 60.4x_1 - 2.30x_2 + 10.7x_3 + 7.93x_{4(II)} -$ $33.4x_1^2 + 2.06x_2^2 + 2.06x_{4(II)}^2 - 0.688x_1x_2 -$ $0.688x_1x_3 - 5.44{,}6x_1x_{4(II)} + 0.688x_3x_{4(II)}$	(7.27)
SP + MK	$F = 120.7 - 46.2x_1 - 6.10x_2 + 7.73x_3 - 5.4x_{4(III)} -$ $29.3x_1^2 - 1.28x_2^2 + 4.22x_3^2 - 8.28x_{4(III)}^2 + 5.125x_1x_2 +$ $1.0x_1x_3 + 1.125x_2x_3 + 5.44x_2x_{4(III)}$	(7.28)

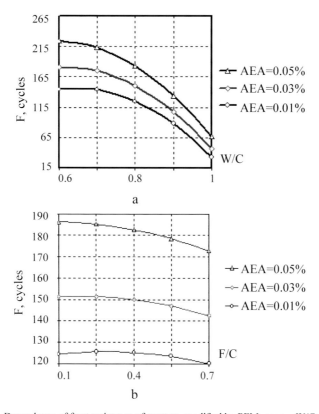

Fig. 7.15 Dependence of frost resistance of mortars, modified by PFM$_1$ on: a – W/C; b – F/C.

mortars with a minimum content of organic additives and low cement consumption (W/C = 1.0), frost resistance was within the range of 25 ... 75. Lower frost resistance has mortar with the highest AGD - cement ratio.

To explain the effect of all investigated additives on the frost resistance of mortars, compressive method was used to determine the volume of entrained air for the characteristic compositions. The results of the measurements are given in Table 7.13.

From the data obtained it follows that in the filled mortar mixtures, which contain 0.7% of SP, air content reaches 2%. This corresponds to the data of V.S. Ramachandran and V.M. Malhotry that in typical mixtures with superplasticizers on the basis of naphthalenesulphonic acid remains 1 ... 3% of air [84].

Approximately the same amount of air is entrained in a mortar mixture with MK. The lowest amount of entrained air was found in mixtures with the maximum content of AGD. Reduced air entraining in mixtures with a high disperse materials content corresponds to the known theoretical concepts [86], according to which wetting of highly dispersed powders requires a significant amount of water, which as a result can no longer perform air entraining and air-containing functions. Additionally, a large number of air entraining additive molecules are adsorbed on highly dispersed

Table 7.13 The content of the entrained air in the modified mortar with characteristic compositions.

Mixture composition						Entrained air volume, %
W/C	F/C	Content of additives				
		SP, %	AEA, %	EC, %	MK, kg/m³	
1.0	0.4	0.3	0.01	-	-	0.7
10	0.7	0.3	0.01	-	-	0.6
1.0	0.4	0.7	0.05	-	-	3.8
1.0	0.4	0.7	-	0.4	-	2.1
1.0	0.1	0.5	-	-	70	1.5
1.0	0.7	0.5	-	-	50	0.9
0.6	0.4	0.3	0.01	-	-	1.4
0.6	0.4	0.7	0.05	-	-	5.3
0.6	0.4	0.7	-	0.4	-	2.0
0.6	0.1	0.5	-	-	70	1.9
0.6	0.7	0.5	-	-	50	1.1

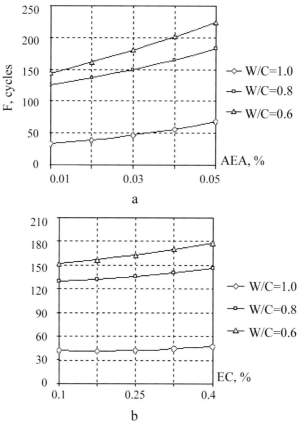

Fig. 7.16 Dependence of modified mortars frost resistance on content of:
a – air entraining admixture; b – ether of cellulose.

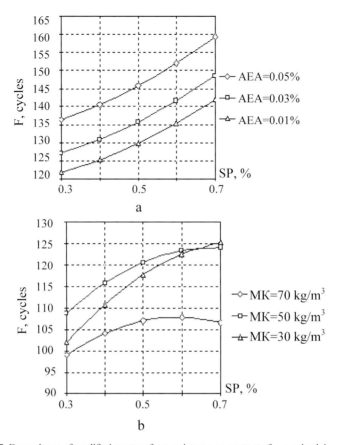

Fig. 7.17 Dependence of modified mortars frost resistance on content of superplasticizer SP.

materials. To compensate significant reduction in air content, an additional amount of air entraining admixtures should be added to the mortar mixture. Adding air entraining admixture raises the content of entrained air in mortars up to 5 ... 7% and increases the frost resistance.

Inclusion of MK in the PFM contributes to reduction in the saturated by water open macro-pores volume. At the same water content in mortar mixtures, the decrease in capillary pores volume may be due to an increase in the cement consumption and its hydration degree. In the considered mortars compositions the total consumptions of cement, AGD and PFM should be taken into account, in other words, the total binder consumption, which interacts with water, should be considered. Mortars with PFM additives, as well as for mortars with MK have a higher ratio between total pore volume and open pore volume, which should also contribute to increased frost resistance.

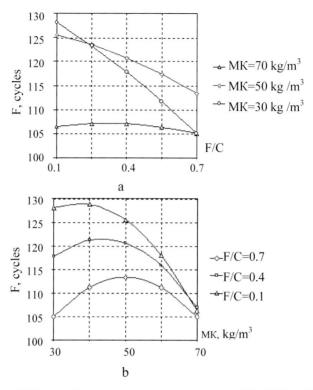

Fig. 7.18 Frost resistance of cement mortars with complex filler (AGD+MK).

7.6 Design of mortars compositions with granite filler

The most important quality indicators for masonry and glue mortars are compressive, tensile and adhesion strength, frost resistance, mortar mixtures fluidity. The main factors of composition, which determine these features, are water – cement, filler – cement and sand – cement ratios, as well as the content of modifying admixtures. The necessary quality indicators of mortars can be obtained at different ratios of these factors. Therefore, when designing the mixture composition there is a need to choose such a ratio of components, which provides the required mortar strength and fluidity as well as its minimum cost. In this case, other normative indicators of the mortar, such as bonding with masonry base, frost resistance, etc., should also be provided.

In the most general form, the problem of mortar composition design can be formulated as a system of equations:

$$
\begin{aligned}
&y_1 = f_1\left(P_1, P_2, \dots P_{n-1}\right) \ge y_1^0; && \left(\text{or} < y_1^0\right) \\
&y_2 = f_2\left(P_1, P_2, \dots P_{n-1}\right) \ge y_2^0; && \left(\text{or} < y_2^0\right) \\
&\dots\dots\dots\dots\dots\dots\dots\dots; \\
&y_{n-1} = f_{n-1}\left(P_1, P_2, \dots P_{n-1}\right) \ge y_{n-1}^0; && \left(\text{or} < y_{n-1}^0\right) \\
&Cr = f\left(P_1, P_2, \dots P_{n-1}\right) \to opt,
\end{aligned}
\tag{7.29}
$$

where y_1^0, y_2^0, y_{n-1}^0 are the required mortar properties indicators;

$P_1, P_2...P_{n-1}$ are the mixture composition factors; Cr is an optimization criterion.

Finding the necessary composition factors by solving the system (7.29) is usually possible using linear equations. When using quadratic equations $y_i = f(P_i)$, it is necessary to determine the possibility of including in the system linear equations $\partial y_i / \partial P_i = 0$, obtained by differentiating the initial parameters by the optimized factors. In some cases, other methods of solving the system (7.29) are possible.

The performed experimental investigations of modified mortars enabled to obtain quadratic mathematical models for the basic normed properties. Design of modified masonry mortar compositions, using AGD as filler, can be performed in the following order:

Set the content of AGD and modifying admixtures close to the optimal values. Analysis of the above experimental results enables to recommend the AGD content, equal to 35 ... 40% of the cement consumption for mortars with compressive strength 10 MPa and higher and 60 ... 70% for mortars with lower strength. To increase the mortars' water-retaining capacity, the AGD content should be at least 100 kg/m³.

The content of superplasticizer SP should be 0.25 ... 0.3% of cement weight for PFM$_1$ modifier, 0.45 ... 0.5% for the PFM$_2$ and 0.35 ... 0.5% for PFM$_3$ modifiers, respectively. Higher values are taken when using very fine sand. The optimal content of air entraining admixture in the PFM$_1$ modifier is 0.04% of the cement weight. The content of Tylose water retaining admixture in PFM$_2$ should be 0.2 ... 0.25% of the cement weight. The recommended content of metakaolin in PFM$_3$ is 5 ... 10% of the cement weight, but not less than 25 kg/m³.

Calculation of the main mortar components is carried out in the following sequence:

1. Find the required W/C from the regression equations that describe the normalized properties of mortars. Such properties are the mortar compressive strength, the strength of its adhesion to the base of the masonry (adhesion strength) and frost resistance. Therefore, mathematical models of strength (7.11), (7.13), (7.15), adhesion strength (7.6), (7.8), (7.10), and frost resistance (7.26 ... 7.28) are used to find W/C. If there are requirements to other of mortar mixtures properties, other regression equations are used. Of all the values obtained, the least W/C, satisfying all the conditions, is selected.
2. By solving the model of the mortar mixture fluidity (7.1 ... 7.3) relative to X$_4$ and switching from coded to natural factors, the of water demand (kg/m³), providing the required mortar mixture fluidity, is obtained.
3. Find the cement consumption (kg/m³):

$$C = \frac{W}{W / C} \tag{7.30}$$

4. According to the known cement consumption, find the contents of AGD and admixtures - components of the polyfunctional modifier - according to the recommendations of paragraph 1.
5. Find the sand consumption (S) from the absolute volumes condition:

$$\frac{C}{\rho_C} + \frac{AGD}{\rho_{AGD}} + \frac{W}{\rho_W} + \frac{S}{\rho_S} + V_a = 1000 \qquad (7.31)$$

where V_a is the volume of air entrained in the mixture, l; ρ_C, ρ_{AGD}, ρ_W, ρ_S are densities of cement, granite dust, water and sand, kg/l.

Therefore

$$S = \left(1000 - \frac{C}{\rho_C} + \frac{AGD}{\rho_{AGD}} + \frac{W}{\rho_W} + V_a \right) \rho_S \qquad (7.32)$$

When minimizing the mortar mixture cost (C_p), the optimization criterion can be expressed by the following linear equation:

$$C_p = P_A A + P_F F + P_C C + P_S S \qquad (7.33)$$

where P_o, P_F, P_C, P_S are prices of modifying admixture, filler, cement and sand, A, F, C and S their contents, respectively.

References

[1] Dvorkin L., Dvorkin O. and Ribakov Y. 2016. *Construction materials based on industrial waste products*, Nova Science Publishers, New York, 231 p.

[2] Chun Y.M., Claisse P., Naik T.R. and Ganjian E. 2007. Sustainable Construction Materials and Technologies. CRC Press, Boca Raton, 816 p.

[3] Khatib J.M. 2016. Sustainability of Construction Materials. 2 edition, Woodhead Publishing, Sawston 740 p.

[4] AalbersTh.G., Goumans J.J.J.M. and van der Sloot H.A. 1991. *Waste Materials in Construction*, Elsevier Science, Amsterdam, 671 p.

[5] Collins R.J. and Ciesielski S.K. 1994. *Recycling and use of waste materials and by-products in highway construction*. Transport Research Board, Washington D.C., 84 p.

[6] Siddique R. 2014. Utilization of industrial by-products in concrete *Procedia Engineering*, 95, Elseiver Science, Amsterdam, 335–347 p.

[7] Siddique R. 2008. *Waste materials and by-products in concrete*Springer, Berlin, 414 p.

[8] Woolley G.R., Goumans, J.J.J.M. and Wainwright P.J. 2000. *Waste materials in construction: science and engineering of recycling for environmental protection*. Elsevier Science, Amsterdam, 1064 p.

[9] Dvorkin L. and Dvorkin O. 2007. Building *materials based on industrial wastes.*Rostov-on-Don, Fenix, 363 p. (in Russian).

[10] Dvorkin L., Pushkareva K., Dvorkin O., Kochevih M., Mohort M. and Bezsmertnih M. 2009.*Using technogenic products in construction*. Rivne, 339 p. (in Ukrainian).

[11] Bozhenov P. 1994. Complex *using mineral raw materials and ecology*. ASV, Moscow, 264 p. (in Russian).

[12] Dvorkin L. 1985. Reducing *cement and energy consumption in precast reinforced concrete production*. Vyshcha Shkola, Kiev, 99 p. (in Russian).

[13] Dvorkin L. and Dvorkin O. 2006. *Basics of concrete science*. Stroybeton, St. Petersburg, 686 p. (in Russian).

[14] Dvorkin L. and Dvorkin O. 2006. Basics of concrete science: optimum design of concrete mixtures. Amazon. (Kindle edition)/(e-book), Stroi-Beton, S-Peterburg, 382 p.

[15] Itskovich S.M., Chumakov L.D. and Bazhenov Y.M. 1991. *Technology of concrete aggregates*. Vysshaya shkola, Moscow, 272 p. (in Russian).

[16] Bazhenov Y.M., Gorchakov G.I., Alimov L.A. and Voronin V.V. 1978. *Obtaining concrete with given properties*. Stroyizdat, Moscow, 54 p. (in Russian).

[17] Berg O.V., Shcherbakov E.N. and Pisanko G.N. 1971. *High strength concrete*. Stroyizdat, Moscow, 208 p. (in Russian).

[18] Dvorkin L.I., Solomatov V.I., Vurovoy V.N. and Chudnovskiy S.M. 1991. *Cement concrete with mineral fillers*. Budivelnyk, Kiev, 137 p. (in Russian).

[19] Dvorkin L. and Dvorkin O. 2011. *Building mineral binding materials*. Infra-Engineering, Moscow, 544 p. (in Russian).

[20] Kosmatka S.H. and Wilson M.L. *Design and control of concrete mixtures*. 2011. 15th edition, Portland Cement Association, Skokie, 460 p.

[21] ACI Manual of Concrete Practice. 1980 Part 2; American Concrete Institute, Detroit, 150 p.

[22] ASTM C33 / C33M-16e1, Standard Specification for Concrete Aggregates, ASTM International, West Conshohocken, PA, 2016, HYPERLINK "http://www.astm.org" www.astm.org

[23] Dvorkin L.I., Dvorkin O.L. and Korneychuk Yu.A. 1995. *Effective cement-ash concrete*. Eden Publishers, Rovno, 195 p. (in Russian).

[24] Stork Y. 1971. *Theory of concrete mixture composition*. Stroyizdat, At. Petersburg, 238 p. (in Russian).

[25] Hewlett P.C. 2004. *Lea's Chemistry of cement and concrete*. 4th edition, Butterworth-Heinemann, Oxford, 1092 p.

[26] Ramachandran V.S., Feldman R.F. and Beaudoin J.J. 1981. *Concrete science: treatise on current research*. Heyden, London.

[27] Neville A.M. *Properties of concrete*. 1996. 4th edition, Wiley & Sons, New York, 844 p.

[28] Ahverdov I.N. 1981. *Basis of concrete physics*. Stroyizdat, Moscow. 464 p. (in Russian).

[29] Powers T. 1958. Structure and Physical Properties of hardened Portland cement paste. *J. Amer. Ceram. Soc.*, 41, 18–26 p.

[30] Sheykin A.E., Chechovskiy Yu.V. and Brusser M.I.1979. *Structure and properties of cement concretes*. Stroyizdat, Moscow, 344 p. (in Russian).

[31] Bazhenov Y.M. 1987.*Concrete technology*. Vysshaya shkola, Moscow, 449 p. (in Russian).

[32] Taylor H.F.W. 1990. *Cement chemistry*. Academic Press, London, 475 p.

[33] Mchedlov–Petrosyan O.P. 1988. *Chemistry of non-organic construction materials*. Stroyizdat, Moscow, 304 p. (in Russian).

[34] Powers T. 1956. The physical structure of Cement and Concrete, *Cement and Lime Manufacture*. 29(2): 270.

[35] Shmigalskiy V.N. 1981. *Optimization of cement concrete compositions*. Shtinca, Chisinau, 123 p. (in Russian).

[36] Ahverdov I.N. 1991. *Theoretical basics of concrete science*, Vysshaya Shkola, Minsk, 188 p. (in Russian).

[37] Gorchakov G.I. 1976. *Composition, structure and properties of cement concrete*. Stroyizdat, Moscow, 145 p.

[38] Dvorkin L. 1981. *Optimal concrete composition design*, Vyshcha Shkola, Lviv, 159 p. (in Ukrainian).

[39] Malhotra V.M. and Mehta P.K. 2002. *High-performance fly ash concrete. Supplementary cementing materials for sustainable development*, Supplementary Cementing Materials for Sustainable Development, Inc.,Ottawa, 101 p.

[40] Malhotra V.M. and Mehta P.K. 2005. *High-performance high-volume fly ash concrete*. Suppementary Cementing Materials for Sustainable Development, Inc.,Ottawa, 120 p.

[41] Wesche K. 1991. *Fly ash concrete: properties and performance*. E&FN Spon, London, 356 p.

[42] Batrakov V.G. 1990. *Modified concretes.* Stroyizdat, Moscow, 396 p. (in Russian).

[43] Dvorkin O. 2003. Concrete with high crack resistance, *Concrete and reinforced concrete in Ukraine*, No. 4, 2003, 2–5 p.(in Ukrainian).

[44] Massazza F. 1976. Chemistry of pozzolanic additives and mixed cements. *Proceedings of the Sixth International Congress on Chemistry of Cement*,Stroyizdat, Moscow, 209–221p.

[45] Moskvin V.M, Ivanov F.M, Alekseev S.K. and Guzeev E.A. 1980. *Corrosion of concrete and reinforced concrete, methods of their protection*, Stroyizdat, Moscow, 536 p.

[46] Dvorkin L., Nwaubani S. and Dvorkin O. 2011. *Construction materials.* Nova Science Publishers, New York, 409 p.

[47] Fedosov S., Akulova M. and Krasnov A. 2008. *High-strength fine grain concrete.* Russia, Ivanovo, 196 p. (in Russian).

[48] Lvovich K. 2007. *Sandy concrete and its using in construction.* Stroi-Beton, S-Peterburg, 250 p. (in Russian).

[49] Sizov V.P. 1980. *Design of normal-weight concrete compositions.* Stroyizdat, Moscow, 144 p. (in Russian).

[50] Dvorkin L., Dvorkin O. and Ribakov Y. 2012. *Mathematical experiments planning in concrete technology.* Nova Science Publishers, New York, 173 p.

[51] Voznesenskiy V.A., Lyashenko T.V. and Ogarkov B.L. 1989. *Numerical methods for solving problems of construction technology using computers*, Vyshcha Shkola, Kiev, 328 p. (in Russian).

[52] Dvorkin L.I. 1981. *Optimal design of concrete compositions.* Vyshcha Shkola, Lvov, 159 p. (in Russian).

[53] Akhnazarova S.L. and Kafarov V.V. 1985. *Methods for optimizing the experiment in chemical technology.* High School, Moscow, 327 p.

[54] Ahverdov I.N. 1961. *High strength concrete.*Stroyizdat, Moscow, 163 p. (in Russian).

[55] Rybiev I.A. and Suleymanov F.G. 1989. *Optimization of concrete compositions based on artificial building conglomerates (ABC) theory and using computers.* VZISI, Moscow, 110 p. (in Russian).

[56] Grushko I.M., Ilin A.G. and Chihladze E.D. 1986. *Increasing concrete strength and endurance.* Vyshcha Shkola, Kharkov, 149 p. (in Russian).

[57] Leshchinsky, M.Yu. 1987. The application of fly ash in concrete and reinforced concrete, *Construction materials*, No. 1, 19–21 p.

[58] Aitcin P.C. 1998. *High-performance concrete.* (Modern Concrete Technology), E & FN Spon, London, 591 p.

[59] Dvorkin L., Bezusyak A., Lushnikova N. and Ribakov Y. 2012. Using mathematical modeling for design of self-compacting high strength concrete with metakaolin admixture. *Construction and Building Materials*, 37: 851–864.

[60] Soroker V.I. and Dovzhik V.G. 1964. *Stiff concrete mixtures in precast reinforced concrete production.* Moscow, Stroyizdat, Moscow, 206 p. (in Russian).

[61] Gambhir M.L. 2004. *Concrete technology 3rd edition.* Tata McGraw-Hill Education, New Delhi, 658 p.

[62] Mikhailov N.V. 1961. Basic principles of the new technology of concrete and reinforced concrete. Stroyizdat, Moscow, 52 p. (in Russian).

[63] Polak A.F. 1976. *Hardening of monomineral binding substances.* Stroyizdat, Moscow. 208 p. (in Russian).

[64] Dzenis V.V. and Lapsa V.Kh. 1971. *Ultrasonic inspection of hardening concrete*. Stroyizdat, Leningrad, 112 p. (in Russian).

[65] Lyse I. 1932. Tests on consistency and strength of concrete having constant water content. *Proceedings of the ASTM, 32,* Part II, 629–636 p.

[66] Ramachandran V.S. 1997. *Concrete admixture handbook. 2nd edition: properties, science and technology*. William Andrew, Norwich, 1183 p.

[67] Mironov S.A. and Malinina L.A. 1964. *Accelerating concrete hardening.* Stroyizdat, 346 p. (in Russian).

[68] Kaiser L.A. and Chehova R.S. 1972. *Rational using of cements for manufacturing precast reinforced concrete elements*. Stroyizdat, Moscow, 80 p. (in Russian).

[69] Shpinova P. 1981. *Physical-chemical basis for the cement stone structure formation.* Vishcha shkola, Lviv, 157 p. (in Russian).

[70] Zhitkevich N. 1912. *Concrete and concrete works*. St. Petersburg, 524 p. (in Russian).

[71] Itskovich S. 1977. *Large-porous concrete (Technology and properties)*. Stroyizdat, Moscow, 119 p. (in Russian).

[72] Aitcin P.C. 2014. *Binders for Durable and Sustainable Concrete*. CRC Press, Boca Raton, 528 p.

[73] Runova R. and Nosovsky Y. 2007. *Technology of modified constructing mortars.* KNUBA Publications, Kiev, 256 p. (in Ukrainian).

[74] Karapuzov E.K., Lutz G., Gerold H., Tolmachev N.G. and Spektor Y.P. 2000. *Dry building mixtures.* Technique, Kiev, 226 p. (in Russian).

[75] Gornostaeva T. 2004. Active filler for concrete, *Building expert*, 14 (177): 3–5 (in Russian).

[76] Artamonov V., Vorobiev V. and Svitov V. 2003. Experience in processing crushing siftings. *Building materials*, 6: 28–29 (in Russian).

[77] Fridrickhsberg D.A. 1986 *Course of colloid chemistry.* Mir Publishers, Moscow, 424 p.

[78] Kalashnikov V. and Demyanova N. 2001. Polymer mineral dry building mixtures. *News of Higher Educational Institutions, Building*, 5: 41–46 (in Russian).

[79] Basin V. 1981. Adhesive *strength*. Chemistry, Moscow, 208 p. (in Russian).

[80] Sychev M. 1987. Prospects of increasing the strength of cement stone. *Cement*. 9: 17–19 (in Russian).

[81] Rebinder P.A. 1966. *Physicochemical Mechanics of Dispersed Structures* Nauka, Moscow, 400 p. (in Russian).

[82] Uriev N. and Dubinin I. 1980. *Colloid-cement mortars*. Stroyizdat, St. Petersburg, 192 p. (in Russian).

[83] Ohama Y. 1978. Development of concrete—polymer materials in Japan, *Proceedings of the Second international congress on polymers in concrete*. Austin, 128–135 p.

[84] Benstein, I.I., Butt Yu. M. and Timashev V. 1971. Crystallization of hydrated neoplasms of cement stone on a carbonate substrate. Works of MHTI, Moscow, 68: 16–22 (in Russian).

[85] Cherkinsky Yu. S. and Slipchenko G. 1976. Hydration hardening of cement in the presence of polymers. *Sixth International Congress on Cement Chemistry,* Vol. 3, Stroyizdat, Moscow, 305–307 p. (in Russian).

[86] Gorchakov G., Kapkin M., Skramtaev B. 1965. *Increase of frost resistance of concrete in constructions of industrial and hydraulic structures*. Stroyizdat, Moscow, 196 p. (in Russian).

Index